Full Stack Django and React

Get hands-on experience in full-stack web development with Python, React, and AWS

Kolawole Mangabo

BIRMINGHAM—MUMBAI

Full Stack Django and React

Group Product Manager: Pavan Ramchandani
Publishing Product Manager: Kushal Dave
Senior Editor: Mark D'Souza
Senior Content Development Editor: Rakhi Patel
Technical Editor: Joseph Aloocaran
Copy Editor: Safis Editing
Project Coordinator: Aishwarya Mohan
Proofreader: Safis Editing
Indexer: Rekha Nair
Production Designer: Aparna Bhagat
Marketing Coordinator: Anamika Singh

First published: February 2023

Production reference: 1200123

Published by Packt Publishing Ltd.
Livery Place
35 Livery Street
Birmingham
B3 2PB, UK.

ISBN 978-1-80324-297-2

`www.packtpub.com`

To my mother, Fleur, and to the memory of my father, Idoumangoye, for their sacrifices and for exemplifying the power of determination. To my uncles, aunties, and grandparents, Prince, Fanick, Carine, Ghislain, Deo Gratias, and Virginie for being my biggest influencers throughout my life journey. To my friends and colleagues, Corentin, Kevin, Ruben, Lewis, and Celda for helping me become a better writer and learner.

– Kolawole Mangabo

Contributors

About the author

Kolawole Mangabo is a software engineer, currently working as a frontend engineer, while regularly using React and Django to build web applications and backends of tech products in various industries, such as foodtech, fintech, and telecom. His goal in a team is to always build products that provide pixel-perfect, performant experiences while adjusting to business and user needs. When he is not coding, he spends most of his time writing and publishing articles on various websites on topics such as software architecture, testing, full stack development, and developer experience.

About the reviewers

Eleke Great (BEng) is a senior software developer with hands-on Django REST framework and ReactJS experience, creating microservices and dynamic UI for high-profile organizations and start-up companies in the US and throughout South America. He has been a Packt Publishing technical reviewer on another of their book projects, *FastAPI React and MongoDB (FARM Stack)*. His other expertise is Next.js, Selenium, Google Cloud Platform, CI/CD, MongoDB, PostgreSQL databases, system design, and database architecture. He is also a full stack blockchain developer with hands-on experience in Web3 with Solidity, Web4, and Web5 with Rust on Solana.

This book is a complete zero-to-hero guide to the DRP stack. The author dedicated time to breaking down Django REST framework and ReactJS and creating production-level, maintainable web applications so that someone from a non-programming background can understand it all and get up and running.

Okere Chinedu is a full stack software engineer from Nigeria, with 4 years of experience. He has worked for big tech companies as well as start-ups, located in the United States, the Middle East, and Africa. He has worked on multiple applications and written many tech articles. He has extensive knowledge of working with JavaScript, React, Node.js, and Python. When he isn't writing code, he is either reading articles, writing one, sharing memes, or playing *Call of Duty*.

Table of Contents

3

Social Media Post Management 51

4

Adding Comments to Social Media Posts 77

5

Testing the REST API 93

Part 2: Building a Reactive UI with React

6

Creating a Project with React 121

7

Building Login and Registration Forms 149

8

Social Media Posts 179

9

Post Comments 213

10

User Profiles 241

11

12

13

Dockerizing the Django Project 315

14

Automating Deployment on AWS 331

15

Deploying Our React App on AWS 345

16

Preface

Getting started with full stack development using Python or JavaScript can be daunting, mainly if you are a developer already using one of these languages and want to add a second language to your set of skills. If you are a developer already working with Django or React, or a developer with knowledge in Python or JavaScript and you want to learn how to build a full stack application from scratch with features such as authentication, CRUD operations, and a lot more, but you are also looking to learn how to deploy web applications on AWS using Docker, this book covers everything you need.

This book will help you to discover the full potential practices while combining the dual power of the two most popular frameworks – React and Django. We will build full stack applications including a RESTful API in the backend and an intuitive frontend while exploring the advanced features of both frameworks. We will start building a social media web application called Postagram from scratch while covering the important concepts, techniques, and best practices for end-to-end development.

We will see how the dynamic functionality of the React framework can be used to build your frontend systems and how the ORM layer of Django helps to simplify a database, which in turn boosts the development process of building a backend to build full stack applications.

By the end of the book, you will be able to create a dynamic full stack app starting from scratch on your own.

Who this book is for

This book is for Python developers who are familiar with Django but don't know where to start when it comes to building a full stack application – more precisely, building a RESTful API. You will also find this book useful if you are a frontend developer with knowledge of JavaScript and looking to learn full stack development. If you are also an experienced full stack developer working with different technologies and you are looking to explore and learn new ones, this book is written for you.

What this book covers

Chapter 1, Creating a Django Project, shows how to create a Django project and make the required configurations with a database server.

Chapter 2, Authentication and Authorization Using JWTs, explains how to implement an authentication system using JSON Web Tokens and how to write custom permissions.

Chapter 3, Social Media Post Management, shows how to implement complex CRUD operations using serializers and ViewSets.

Chapter 4, Adding Comments to Social Media Posts, shows how to add comments to posts using database relations, serializers, and viewsets.

Chapter 5, Testing the REST API, introduces you to testing with Django and Pytest.

Chapter 6, Creating a Project with React, explains how to create a React project while configuring a good environment for development.

Chapter 7, Building Registration and Login Forms, explains how to implement authentication forms and logic on the frontend side of a full stack application.

Chapter 8, Social Media Posts, shows how to implement CRUD operations on the React frontend for social media posts.

Chapter 9, Post Comments, shows how to implement CRUD operations on the React frontend for social media comments.

Chapter 10, User Profiles, explains how to implement CRUD operations on the React frontend concerning profiles and how to upload an image.

Chapter 11, Effective UI Testing for React Components, introduces you to component testing using Jest and the React Testing Library.

Chapter 12, Deployment Basics – Git, GitHub, and AWS, introduces DevOps tools and terms and how to deploy a Django application directly on AWS EC2.

Chapter 13, Dockerizing the Django Project, shows how to dockerize a Django application using Docker and Docker Compose.

Chapter 14, Automating Deployment on AWS, shows how to deploy a dockerized application on EC2 using GitHub Actions.

Chapter 15, Deploying Our React App on AWS, demonstrates how to deploy a React application on AWS S3 and automate the deployment using GitHub Actions.

Chapter 16, Performance, Optimization, and Security, shows you how to optimize your application using webpack, optimize database queries, and enhance the backend security.

To get the most out of this book

You will need Python 3.8+, Node.js 16+, and Docker installed on your machine for this book. All code and examples in this book are tested using Django 4.1 and React 18 on Ubuntu. When installing any React or JavaScript libraries, ensure that you have the latest installation command (`npm`, `yarn`, and pnpm) from their documentation, and check whether there are any major changes related to the version used in this book.

Software/hardware covered in the book	Operating system requirements
Python	Windows, macOS, or Linux
JavaScript	Windows, macOS, or Linux
PostgreSQL	Windows, macOS, or Linux
Django	Windows, macOS, or Linux
React	Windows, macOS, or Linux
Docker	Windows, macOS, or Linux

If you are using the digital version of this book, we advise you to type the code yourself or access the code from the book's GitHub repository (a link is available in the next section). Doing so will help you avoid any potential errors related to the copying and pasting of code.

Download the example code files

You can download the example code files for this book from GitHub at https://github.com/PacktPublishing/Full-stack-Django-and-React. If there's an update to the code, it will be updated in the GitHub repository.

We also have other code bundles from our rich catalog of books and videos available at https://github.com/PacktPublishing/. Check them out!

Download the color images

We also provide a PDF file that has color images of the screenshots and diagrams used in this book. You can download it here: https://packt.link/jdEHp.

Conventions used

There are a number of text conventions used throughout this book.

Code in text: Indicates code words in text, database table names, folder names, filenames, file extensions, pathnames, dummy URLs, user input, and Twitter handles. Here is an example: "Once the package is installed, create a new file called pytest.ini at the root of the Django project."

A block of code is set as follows:

```
>>> comment = Comment.objects.create(**comment_data)
>>> comment
```

```
<Comment: Dingo Dog>
>>> comment.body
'A comment.'
```

When we wish to draw your attention to a particular part of a code block, the relevant lines or items are set in bold:

```
ENV = os.environ.get("ENV")

# SECURITY WARNING: keep the secret key used in production
secret!
SECRET_KEY = os.environ.get(
    "SECRET_KEY", default="qkl+xdr8aimpf-&x(mi7)dwt^-
q77aji#j*d#02-5usa32r9!y"
)

# SECURITY WARNING: don't run with debug turned on in
production!
DEBUG = False if ENV == "PROD" else True

ALLOWED_HOSTS = os.environ.get("DJANGO_ALLOWED_HOSTS",
default="*").split(",")
```

Any command-line input or output is written as follows:

```
pip install drf-nested-routers
```

Bold: Indicates a new term, an important word, or words that you see on screen. For instance, words in menus or dialog boxes appear in **bold**. Here is an example: "Finally, select the **Permissions** tab and select **Bucket Policy**."

> **Tips or important notes**
> Appear like this.

Get in touch

Feedback from our readers is always welcome.

General feedback: If you have questions about any aspect of this book, email us at `customercare@packtpub.com` and mention the book title in the subject of your message.

Errata: Although we have taken every care to ensure the accuracy of our content, mistakes do happen. If you have found a mistake in this book, we would be grateful if you would report this to us. Please visit `www.packtpub.com/support/errata` and fill in the form.

Piracy: If you come across any illegal copies of our works in any form on the internet, we would be grateful if you would provide us with the location address or website name. Please contact us at `copyright@packt.com` with a link to the material.

If you are interested in becoming an author: If there is a topic that you have expertise in and you are interested in either writing or contributing to a book, please visit `authors.packtpub.com`.

Share Your Thoughts

Once you've read *Full Stack Django and React*, we'd love to hear your thoughts! Scan the QR code below to go straight to the Amazon review page for this book and share your feedback.

`https://packt.link/r/1803242973`

Your review is important to us and the tech community and will help us make sure we're delivering excellent quality content.

Download a free PDF copy of this book

Thanks for purchasing this book!

Do you like to read on the go but are unable to carry your print books everywhere? Is your eBook purchase not compatible with the device of your choice?

Don't worry, now with every Packt book you get a DRM-free PDF version of that book at no cost.

Read anywhere, any place, on any device. Search, copy, and paste code from your favorite technical books directly into your application.

The perks don't stop there, you can get exclusive access to discounts, newsletters, and great free content in your inbox daily

Follow these simple steps to get the benefits:

1. Scan the QR code or visit the link below

`https://packt.link/free-ebook/9781803242972`

2. Submit your proof of purchase
3. That's it! We'll send your free PDF and other benefits to your email directly

Part 1: Technical Background

In this part of the book, you will learn how to build a REST API with Django and Django REST. This part provides the knowledge required to connect Django to a PostgreSQL database, add authentication using JSON Web Tokens, create RESTful resources supporting complex CRUD operations, and add tests to a Django application. We will specifically build the backend of a social media web application called Postagram with the most common features of a social media application, such as post management, comment management, and post likes.

This section comprises the following chapters:

- *Chapter 1, Creating a Django Project*
- *Chapter 2, Authentication and Authorization Using JWTs*
- *Chapter 3, Social Media Post Management*
- *Chapter 4, Adding Comments to Social Media Posts*
- *Chapter 5, Testing the REST API*

1
Creating a Django Project

Django is one of the most famous backend frameworks written in Python and is often used to build simple or complex web applications. As for **React**, it's one of the most widely used JavaScript libraries to create reactive and powerful user interfaces. In this chapter, we'll focus on Django first.

In this chapter, we'll briefly explain **software development** and, in particular, **backend development** in the context of what we'll be building: a social network web application with Django and React. We'll also talk about the most common tools used for backend development in **Python** – here in Django. Then, we will create a Django project and explain the most important parts of a Django project. After that, we'll connect **PostgreSQL** to the Django project.

By the end of this chapter, you'll understand concepts such as software development, frontend development, and backend development. You'll also learn how to create a project in Django and start a server.

In this chapter, we'll be covering the following topics:

- An overview of software development
- Understanding backend development
- What is an API?
- What is Django?
- Setting up the work environment
- Configuring the database

An overview of software development

Software development is a complex process full of many steps and many components. These components ensure that conceiving, specifying, designing, programming, documenting, and testing an application, a framework, or software is respected and well applied.

Generally, the software is made of the following two components:

- The **backend**: This represents what the user can't see; it's composed of the business logic and data manipulation from a database
- The **frontend**: This represents the interface provided to the user to interact with the whole application

The term frontend refers to the elements of a site or application that users see onscreen and with which they will interact. For example, all internet users will see a combination of HTML, CSS, and JavaScript on a website. It is these frontend programming languages that will be interpreted by the browser.

Typically, the frontend consists of HTML, CSS, JavaScript, and jQuery (or other UI libraries or frameworks) used to replicate a design. The design is created by the web designer who will create graphic models with dedicated tools, such as Photoshop or Figma.

Here, we'll focus on web development. Web development is the part of software development focused on building websites and web applications, and the notion of web development relies on a client-server architecture.

The client-server architecture represents an environment in which applications running on a client machine can communicate with other applications installed on a server machine, which provides services or data from a database.

On the web, the client will simply be a browser used to request a page or a resource from a server.

Here's a simple diagram demonstrating this:

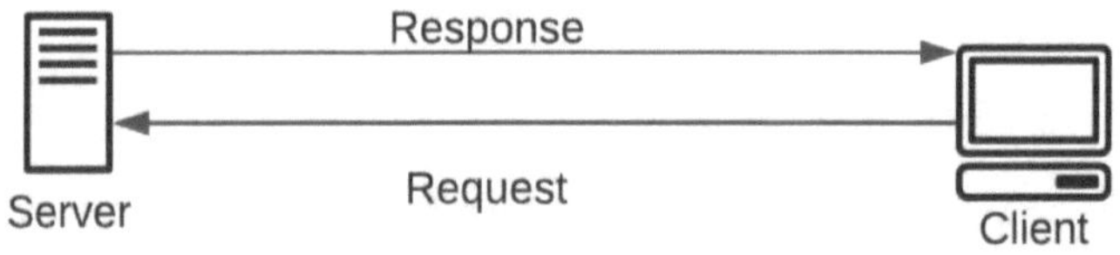

Figure 1.1 – Client-server architecture

Now that we have a better understanding of software development, particularly web development, let's move on to a component of it: backend development.

Understanding backend development

Backend development handles the behind-the-scenes of modern applications. Most of the time, it's made of code that connects to the database, manages user connections, and also powers web applications or the **API**.

The focus of backend development code is more on the business logic. It primarily focuses on how an application works and the functionality and logic powering the application.

For example, let's talk about a web application built to manage books. Let's suppose that the application is connected to an SQL database.

Whatever language is used to build the application and the structure, here are some requirements that represent the business logic and that primarily depend on the backend rather than the frontend:

- **Adding a book (only for admin)**: This supposes that the client (frontend) should be able to make a request to an API powered using whatever language is built for the backend, containing the data needed to create a new entry in the database that represents a book. This action is only available to admins.
- **Listing all books**: This supposes that the client should also be able to make a request to the API, and this API should send as a response a list of all the books in JSON/XML format.

Just by taking a look at these two requirements, we can quickly understand that the frontend will just be the interface through which to request these actions. However, the backend will (taking the first requirement as an example) make sure that the incoming request is possible (checking for permissions such as whether the user making the request is really an admin) and that the data in the request is valid – only after that can data be safely registered in the database. Backend developers use programming languages such as Python, PHP, and Ruby to set up and configure the server. These tools will allow them to store, process, and modify information. To make these programming languages even more practical, developers will improve them with frameworks such as Symfony, Ruby on Rails, CakePHP, or CodeIgniter. These tools will make development faster and more secure. They must then ensure that these tools are always up to date and facilitate the maintenance required.

A backend developer is therefore responsible for creating and managing all the elements invisible to the end user. It is therefore they who are responsible for all the functionalities of the site or the application. They are also responsible for creating the database which will allow, among other things, the information provided by users to be retained. For example, the backend developer will use the databases to find the usernames and passwords that customers have used to connect. It is possible to train for this profession by training in web development or even training in Python.

Responsibilities of backend developers

The backend is typically made of three major parts:

- **Server**: A machine or an application (NGINX) that receives requests
- **Application**: A running application on the server that receives the requests, validates these requests, and sends an appropriate response
- **Database**: Used to store data

Thus, the responsibilities of backend programmers could easily involve writing APIs, writing code to interact with a database, creating modules or libraries, also working on business data and architecture, and much more.

They also have to do the following:

- Coordinate and communicate with frontend developers to transfer data efficiently to the client side of the application
- Collaborate with quality assurance engineers to optimize the server-side processes and also pass some security checks
- Optimize the application when the number of requests or users scales as well
- Analyze the requirements of the project and create a simple structure to handle bugs and errors
- Propose efficient solutions for cloud hosting but also build CI/CD pipelines

The backend architecture actually helps build one of the most common interfaces for consuming data in the software industry: an **Application Programming Interface** (**API**). Let's learn more about the term.

What is an API?

In this book, we'll primarily be building an API – so, what is an API?

Before answering this question, just remember that most of the internet is powered by **Representational State Transfer** (**REST**) or **RESTful APIs**. An API simplifies the way data is exchanged between applications or machines. It consists mainly of two components:

- The technical specification, which describes the data exchange options between the parties, with the specification made in the form of a request for data delivery protocols and data processing
- The software interface (the programming code), which is written to the specification that represents it

For example, if the client side of your application is written in JavaScript and the server side is written in PHP, you'll need to create a web API with PHP (as data comes from the database), which will help you write the rules and routes that will be used to access data.

Web APIs are relatively common and there are different specifications and protocols. The goal of API specification is to standardize—because of different programming languages and different **Operating Systems** (**OSs**)—exchanges between two or more web services. For example, you'll find the following:

- **Remote Procedure Call** (**RPC**): A protocol that can be used by a program to request a service from a program on another computer on a network that it does not need to know the details of. This is sometimes called a function or subroutine call.

- **Simple Object Access Protocol (SOAP)**: An XML-based communication protocol that allows applications to exchange information with each other over HTTP. It therefore allows access to web services and the interoperability of applications across the web. SOAP is a simple and lightweight protocol that relies entirely on established standards such as HTTP and XML. It is portable and therefore independent of any OS and type of computer. SOAP is a non-proprietary specification.
- **REST/RESTful**: A style of architecture for building applications (web, intranet, or web service). This is a set of conventions and best practices to be observed, not a technology in its own right. The REST architecture uses the original specifications of the HTTP protocol, rather than reinventing an overlay (as SOAP or XML-RPC do, for example):
  - **Rule 1**: The URL is a resource identifier
  - **Rule 2**: HTTP verbs are identifiers of operations
  - **Rule 3**: HTTP responses are representations of resources
  - **Rule 4**: Links are relations between resources
  - **Rule 5**: A parameter is an authentication token

In this book, we'll be building REST APIs using Django and **Django REST**, so let's get to know REST a bit better.

Understanding REST APIs

REST is usually the way to go when developers want to build an API. REST is a simple alternative to SOAP and RPC, as it makes it easier to write the logic to access resources; resources here are represented by a unique URL available with one request to this URL.

RESTful APIs use HTTP requests (or methods) to interact with resources:

- `GET`: The most commonly used method in APIs and websites. This method is used to retrieve data from a server at a specified resource. This resource is an endpoint returning an object or a list of objects in JSON or XML most of the time.
- `POST`: The `POST` method is a basic method for requesting information processing from the server. These requests are supposed to bring mechanisms specific to the server into play and cause communications with other modules, or even other servers, to process said data. Therefore, it is quite likely that two identical `POST` requests will receive different or even semantically opposite responses. The data to be processed is specified in the body of the request. The document designated by the request via the page is the resource that must process the data and generate the response.

- `HEAD`: The `HEAD` method is used to query the header of the response, without the file being sent to you immediately. This is useful, for example, if large files need to be transferred: thanks to the `HEAD` request, the client can be informed of the size of the file first and only then decide whether to receive the file.
- `OPTIONS`: This is a diagnostic method, which returns a message that is useful primarily for debugging and the like. This message basically indicates, surprisingly, which HTTP methods are active on the web server. In reality, it's rarely used for legitimate purposes these days, but it does give potential attackers a bit of help – it can be seen as a shortcut to finding another hole.
- `DELETE` and `PUT`: These methods are supposed to allow a document to be uploaded (to the server) or deleted without going through an **File Transfer Protocol** (**FTP**) server or the like. Obviously, this can cause file replacements, and therefore very large security breaches on a server. Therefore, most web servers require a special configuration with a resource or a document responsible for processing these requests. The document referred to by the request is the one to be replaced (or created), and the content of the document is in the body of the request. In theory, URL parameters and the fragment identifier should be prohibited or ignored by the server. In practice, they are generally transmitted to the resource responsible for processing the request.
- `PATCH`: The `PATCH` method of an HTTP request applies partial changes to a resource.
- `TRACE`: The `TRACE` method can be used to trace the path that an HTTP request takes to the server and then to the client.
- `CONNECT`: This method is supposed to be used to request the use of the server as a proxy. Not all servers necessarily implement them.

One interesting benefit is that RESTful systems support different data formats, such as plain text, HTML, YAML, JSON, and XML.

As mentioned previously, in this book, we'll be building REST APIs using Django and Django REST.

What is Django?

Django is an advanced web framework that was first released in 2005. It is written in Python and makes use of the **Model-View-Controller** (**MVC**) architectural pattern. This pattern is commonly defined as follows:

- **Model**: Corresponds to all the data-related logic. It's deeply connected to the database, as it provides the shape of the data but also methods and functions for **Create, Read, Update, and Delete** (**CRUD**) operations.
- **View**: Handles the UI logic of the application.
- **Controller**: Represents a layer between the model and view. Most of the time, controllers interpret the incoming requests from the view, manipulate the data provided by the model component, and interact with the view again to render the final output.

In Django, this will be referred to as the **Model-View-Template** (**MVT**) architecture with the template corresponding to the view and the view here represented by the controller. Here's a simple representation of the MVT architecture:

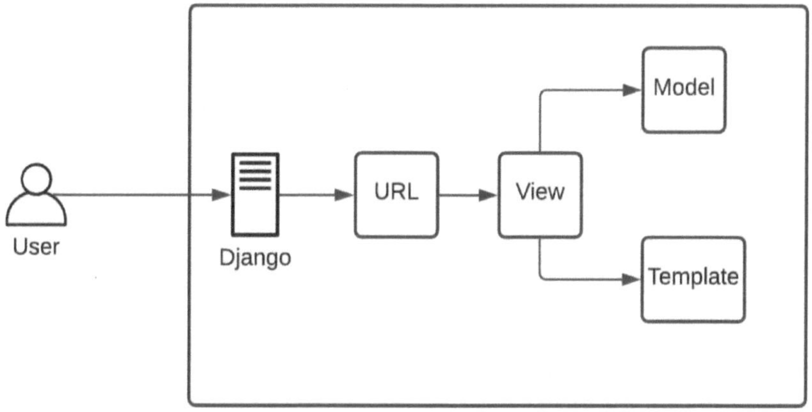

Figure 1.2 – MVT architecture

Django is a web framework that adopts the *Batteries included* approach. When developing a custom web application, Django provides the tools required to speed up the development process. It provides code and tools for common operations such as database manipulation, HTML templates, URL routing, session management, and security.

Django allows developers to build all kinds of web applications (social networks, news sites, and wikis) with all the necessary basics, such as application security, made available from the start to allow developers to fully concentrate on most of their projects. Django provides protection against commons attacks – cross-site scripting, SQL injection, and much more.

Here, we'll also be using **Django REST Framework** (**DRF**). It's the most mature, testable, well-documented, and easily extendable framework, which will help create powerful RESTful APIs when coupled with Django. The combination of Django and the DRF is used by large companies such as Instagram, Mozilla, and even Pinterest.

When this framework is coupled with Django, the view will be replaced by routes or endpoints. We'll discuss this concept later in the book – but why build an API with Django?

It's true that traditional Django supports client languages such as HTML, CSS, and JavaScript. This helps build user interfaces that are served by the server and the performance is always impressive.

However, what if you have many machines that'll access resources on the Django server? It's true that if these machines are running applications based on JavaScript, we can always use the traditional Django way.

What if it's a mobile application? What if it's a service written with PHP?

That's where an API can really be useful. You can have as many machines as you want requesting data from your API without issue, irrespective of the technology or the language used to build the applications that these machines are running.

Now that you have an idea about what Django is, let's set up the working environment and create our first server in Django.

Setting up the work environment

Before starting to work with Django, we must make sure you have a great environment, whatever OS you are using right now.

First of all, make sure you have the latest version of Python installed. For this book, we'll be working with Python 3.10.

If you are using a Windows machine, go to the official download page at `https://www.python.org/downloads/` and download the relevant version.

For Linux users, you can download it using the default repository package download manager.

Creating a virtual environment

Now that we have Python installed, we have to ensure that we have `virtualenv` installed:

```
python3 -m pip install --user virtualenv
```

See the following for Windows users:

```
py -m pip install --user virtualenv
```

Once this is done, we can now create a virtual environment – but why?

There are two types of environments when developing with Python: the global environment and the local environment.

If you just enter `pip` install requests randomly in the terminal, the package will be installed and can be accessed globally: this means accessed anywhere on your machine. Sometimes, you want to isolate the working environment to avoid version conflicts. For example, globally you may be working with Python 3.5, which supports Django 2.x versions. However, for this project, you want to use Python 3.10 and the latest version of Django – here, 4.0. Creating a `virtualenv` environment helps you with that.

Now that we have `virutalenv` installed, we can create and activate the `virtualenv` environment – but before that, create a directory called `django-api`. We'll be building the Python project here.

See the following for Unix or macOS:

```
python3 -m venv venv
```

See the following for Windows:

```
py -m venv venv
```

These preceding commands will create the `venv` directory containing the installed Python packages and the necessary configuration to access these packages when the virtual environment is activated. The next step is to activate the virtual environment. This will help us install the packages we need to start working on.

See the following for Unix or macOS:

```
source venv/bin/activate
```

See the following for Windows:

```
.\venv\Scripts\activate
```

Great! Next, let's install the Django package.

Installing Django

There are two ways to install packages in Python. You can easily just run `pip install package_name`.

Alternatively, you can write the package name with the version in a text file. I'll go with the latter but feel free to use whatever version works for you.

Just understand that there can be some changes between the version and it can affect your project. For more similarities with what we'll be using here, you can also use the latter option.

Great – let's create a file named `requirements.txt` at the root of the `django-api` directory and add the Django package name:

```
Django==4.0
```

Great! Now, run `pip install -r requirements.txt` to install Django.

To make sure everything is working, we'll quickly create a simple project.

Creating a sample project

To create a new project, we'll use the `django-admin` command. It comes with options we can use to create projects in Django:

```
django-admin startproject CoreRoot .
```

Don't forget to add the . dot at the end of this command. This will actually generate all the files in the current directory instead of creating another directory to put all the files in.

You should have a structure of a file such as this:

Figure 1.3 – File structure

Before starting the server, let's run the migrations:

```
python manage.py migrate
```

You'll have a similar output:

```
Operations to perform:
  Apply all migrations: admin, auth, contenttypes, sessions
Running migrations:
  Applying contenttypes.0001_initial... OK
  Applying auth.0001_initial... OK
  Applying admin.0001_initial... OK
  Applying admin.0002_logentry_remove_auto_add... OK
  Applying admin.0003_logentry_add_action_flag_choices... OK
  Applying contenttypes.0002_remove_content_type_name... OK
  Applying auth.0002_alter_permission_name_max_length... OK
  Applying auth.0003_alter_user_email_max_length... OK
  Applying auth.0004_alter_user_username_opts... OK
  Applying auth.0005_alter_user_last_login_null... OK
  Applying auth.0006_require_contenttypes_0002... OK
  Applying auth.0007_alter_validators_add_error_messages... OK
  Applying auth.0008_alter_user_username_max_length... OK
  Applying auth.0009_alter_user_last_name_max_length... OK
  Applying auth.0010_alter_group_name_max_length... OK
  Applying auth.0011_update_proxy_permissions... OK
  Applying auth.0012_alter_user_first_name_max_length... OK
  Applying sessions.0001_initial... OK
```

Migrations are just a way to propagate changes made to the model in the database schema. As Django also comes with some models (such as the `User` model you can use for authentication), we need to apply these migrations. When we write our own models, we'll also be creating migrations files and migrating them. Django has **object-relational mapping** (**ORM**) that automatically handles the interaction with the database for you.

Learning SQL and writing your own queries is quite difficult and demanding when you are new to it. It takes a long time and is quite off-putting. Fortunately, Django provides a system to take advantage of the benefits of an SQL database without having to write even a single SQL query!

This type of system is called ORM. Behind this somewhat barbaric-sounding name hides a simple and very useful operation. When you create a model in your Django application, the framework will automatically create a suitable table in the database that will save the data relating to the model.

No need to write SQL commands here – we'll just write code in Python that will be directly translated into SQL. `python manage.py migrate` will then apply these changes to the database.

Now, run `python manage.py runserver`. You'll see a similar output, and you'll also have your server running at `https://localhost:8000`.

Just hit this URL in your browser and you will see something such as this:

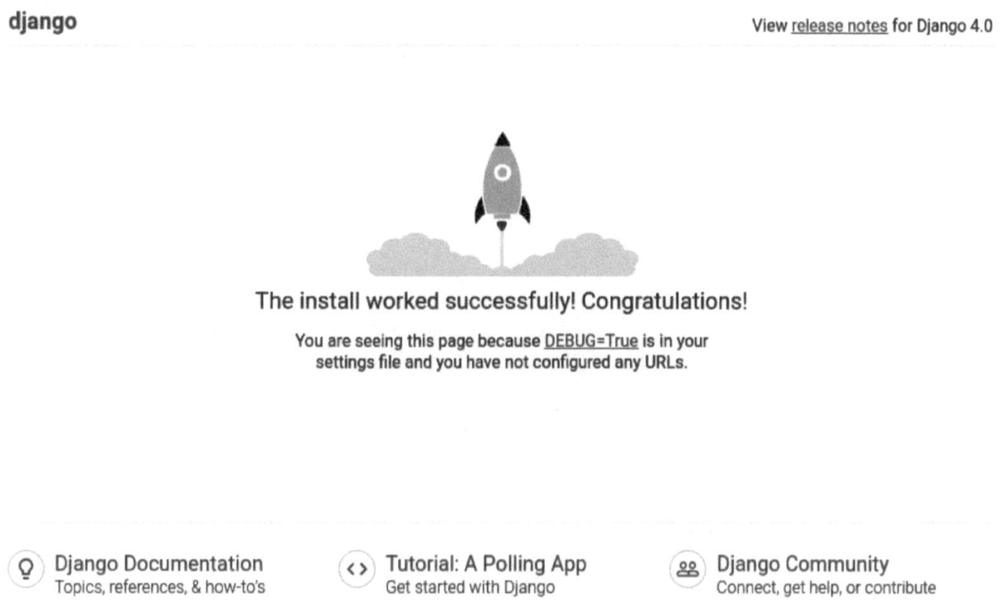

Figure 1.4 – Welcome page of the Django running server

Great – we've just installed Django and started a Django server. Let's talk about the structure of the project.

Discussing the sample project

In the last part, we've briefly talked about how to create a `virtualenv` environment with Python. We've also created a Django project and made it run.

Let's talk quickly about the project.

You may have noticed some files and directories in the `django-api` directory. Well, let's quickly talk about these:

- `manage.py`: This is a utility provided by Django for many different needs. It'll help you create projects and applications, run migrations, start a server, and so on.
- `CoreRoot`: This is the name of the project we've created with the `django-admin` command. It contains files such as the following:
  - `urls.py`: This contains all the URLs that will be used to access resources in the project:

    ```
    from django.contrib import admin
    from django.urls import path
    urlpatterns = [
        path('admin/', admin.site.urls),
    ]
    ```

  - `wsgi.py`: This file is basically used for deployment but also as the default development environment in Django.
  - `asgi.py`: Django also supports running asynchronous codes as an ASGI application.
  - `settings.py`: This contains all the configurations for your Django projects. You can find `SECRET_KEY`, the `INSTALLED_APPS` list, `ALLOWED_HOST`, and so on.

Now that you are familiar with the structure of a Django project, let's see how to configure the project to connect to a database.

Configuring the database

Django, by default, uses **sqlite3** as a database, which is an in-process library that implements a fast self-contained, zero-configuration, serverless, transactional SQL database engine. It's very compact and easy to use and set up. It's ideal if you are looking to quickly save data or for testing. However, it comes with some disadvantages.

First of all, there are no multi-user capabilities, which means that it comes with a lack of granular access control and some security capabilities. This is due to the fact that SQLite reads and writes directly to an ordinary disk file.

For example, in our project, after running the migrations, you'll notice the creation of a new file, `db.sqlite3`. Well, this is our database actually.

We will be replacing it with a more powerful SMDB called **Postgres**.

Postgres configuration

PostgreSQL is one of the world's most advanced enterprise-class open source database management systems, developed and maintained by the PostgreSQL global development group. It's a powerful and highly extensible object-relational SQL database system that comes with interesting features such as the following:

- User-defined types
- Table inheritance
- Asynchronous replication
- Multi-user capabilities

These are the features you will be looking for in a database, mostly when working in a development or production environment.

According to your OS, you can download Postgres versions at `https://www.postgresql.org/download/`. In this book, we are working with PostgreSQL 14.

Once it's done, we'll install a PostgreSQL adapter for Python, **psycopg**:

```
pip install psycopg2-binary
```

Don't forget to add this to the `requirements.txt` file:

```
Django==4.0
psycopg2_binary==2.9.2
```

Great – now that we have the adapter installed, let's quickly create the database we'll use for this project.

For that, we need to connect as a Postgres user in the terminal and then access the `psql` terminal. In that terminal, we can enter SQL commands.

For Linux users, you can log in as follows:

```
sudo su postgres
```

Then, enter `psql`.

Great – let's create the database:

```
CREATE DATABASE coredb;
```

To connect to the database, we need USER with a password:

```
CREATE USER core WITH PASSWORD 'wCh29&HE&T83';
```

It's always a good habit to use strong passwords. You can generate strong passwords at https://passwordsgenerator.net/ – and the next step is to grant access to our database to the new user:

```
GRANT ALL PRIVILEGES ON DATABASE coredb TO core;
```

We are nearly done. We also need to make sure this user can create a database. This will be helpful when we can run tests. To run tests, Django will configure a full environment but will also use a database:

```
ALTER USER core CREATEDB;
```

With that, we are done with the creation of the database. Let's connect this database to our Django project.

Connecting the database

Connecting the database to Django requires some configurations. Then, we have to open the `settings.py` file, look for a database configuration, and then modify it.

In the `settings.py` file, you'll find a similar line:

```
# Database
# https://docs.djangoproject.com/en/4.0/ref          /
settings/#databases
DATABASES = {
    'default': {
        'ENGINE': 'django.db.backends.sqlite3',
        'NAME': BASE_DIR / 'db.sqlite3',
    }
}
```

Great – as you can see, the project is still running on the SQLite3 engine.

Remove this content and replace it with this:

```
DATABASES = {
    'default': {
```

```
        'ENGINE': 'django.db.backends.postgresql_psycopg2',
        'NAME': coredb,
        'USER': 'core',
           'PASSWORD': 'wCh29&HE&T83',
        'HOST': 'localhost',
        'PORT': '5342',
    }
}
```

We've just modified the database engine but also filled in information such as the name of the database, the user, the password, the host, and the port.

The `ENGINE` key for the MySQL database varies. Besides that, there are a few additional keys, such as `USER`, `PASSWORD`, `HOST`, and `PORT`:

- `NAME`: This key stores the name of your MySQL database
- `USER`: This key stores the username of the MySQL account to which the MySQL database will be connected
- `PASSWORD`: This key stores the password for this MySQL account
- `HOST`: This key stores the IP address at which your MySQL database is hosted
- `PORT`: This key stores the port number on which your MySQL database is hosted

The configuration is done. Let's run the migrations and see whether everything works okay:

```
python manage.py migrate
```

You will get a similar output in the terminal:

```
Operations to perform:
  Apply all migrations: admin, auth, contenttypes, sessions
Running migrations:
  Applying contenttypes.0001_initial... OK
  Applying auth.0001_initial... OK
  Applying admin.0001_initial... OK
  Applying admin.0002_logentry_remove_auto_add... OK
  Applying admin.0003_logentry_add_action_flag_choices... OK
  Applying contenttypes.0002_remove_content_type_name... OK
  Applying auth.0002_alter_permission_name_max_length... OK
  Applying auth.0003_alter_user_email_max_length... OK
```

```
Applying auth.0004_alter_user_username_opts... OK
Applying auth.0005_alter_user_last_login_null... OK
Applying auth.0006_require_contenttypes_0002... OK
Applying auth.0007_alter_validators_add_error_messages... OK
Applying auth.0008_alter_user_username_max_length... OK
Applying auth.0009_alter_user_last_name_max_length... OK
Applying auth.0010_alter_group_name_max_length... OK
Applying auth.0011_update_proxy_permissions... OK
Applying auth.0012_alter_user_first_name_max_length... OK
Applying sessions.0001_initial... OK
```

Great! We've just configured Django with PostgreSQL.

Installing an HTTP request client

When developing an API as a backend developer, it's a good habit to have an API client to test your API and make sure it behaves as you needed. API clients are packages, or libraries to send HTTP requests to an API. A great majority supports features such as SSL checking, authentication, and header modification. In this book, we'll be working with Insomnia. It's lightweight and simple to use and customize.

To download a version of Insomnia that suits your OS, go to the following page: `https://insomnia.rest/download`.

Summary

In this chapter, we explored the world of backend development to clarify the roles and responsibilities of a backend developer. We also talked about APIs, mostly REST APIs, which will be built in this book. We've also had a brief introduction to Django, the MVT architecture used by the framework, and connected a PostgreSQL database to the Django project.

In the next chapter, we will dig deeper into Django by creating our first models, tests, and endpoints.

Questions

1. What is a REST API?
2. What is Django?
3. How to create a Django project?
4. What are migrations?
5. What is a virtual environment in Python?

2

Authentication and Authorization using JWTs

In this chapter, we'll dive deeper into Django and its architecture. We'll be working with **models**, **serializers**, and **viewsets** to create an API that can receive HTTP requests as well as return a response. This will be done by building an authentication and authorization system using **JSON Web Tokens (JWTs)** to allow users to create an account, log in, and log out.

By the end of this chapter, you'll be able to create Django models, write Django serializers and validation, write viewsets to handle your API requests, expose your viewsets via the Django REST routers, create an authentication and authorization system based on JWTs, and understand what a JWT is and how it helps with authentication and permissions.

We will be covering the following topics in this chapter:

- Understanding JWTs
- Organizing a project
- Creating a user model
- Writing the user registration feature
- Adding the login feature
- Refresh logic

Technical requirements

For this chapter, you'll need to have Insomnia installed on your machine to make requests to the API we'll be building.

You can also find the code of this chapter at `https://github.com/PacktPublishing/Full-stack-Django-and-React/tree/chap2`.

Understanding JWTs

Before writing the authentication feature, let's explain what a JWT is. As mentioned earlier, **JWT** stands for **JSON Web Token**. It's one of the most used means of authentication in web applications but also helps with authorization and information exchanges.

According to RFC 7519, a JWT is a JSON object defined as a safe way of transmitting information between two parties. Information transmitted by JWT is digitally signed so it can be verified and trusted.

A JWT contains three parts—a header (`x`), a payload (`y`), and a signature (`z`)—that are separated by a dot:

```
xxxxx.yyyyy.zzzzz
```

- **Header**

  The header of the JWT consists of two parts: the type of token and the signing algorithm being used. The signing algorithm is used to ensure that the message is authentic and not altered.

  Here's an example of a header:

  ```
  {
        "alg": "RSA",
        "typ": "JWT"
  }
  ```

  Signing algorithms are algorithms used to sign tokens issued for your application or API.

- **Payload**

  The payload is the second part that contains the claims. According to the official JWT documentation (`https://jwt.io/introduction`), claims are statements about an entity (typically, the user) and additional data.

  Here's an example of a payload:

  ```
  {
     "id": "d1397699-f37b-4de0-8e00-948fa8e9bf2c",
     "name": "John Doe",
     "admin": true
  }
  ```

  In the preceding example, we have three claims: the ID of the user, the name of the user, and also a Boolean for the type of user.

- **Signature**

  The signature of a JWT is the encoded header, the encoded payload plus a secret, and an algorithm specified in the header, all of them combined and signed.

  For example, it's possible to create a signature the following way using the RSA algorithm:

  ```
  RSA(
    base64UrlEncode(header) + "." +
    base64UrlEncode(payload),
    secret)
  ```

  The role of the signature is to track whether information has been changed.

But how are JWTs actually used in authentication?

Understanding how JWTs are used in authentication

Each time a user successfully logs in, a JWT is created and returned. The JWT will be represented as credentials used to access protected resources. The fact that it's possible to store data in a JWT makes it vulnerable. That's why you should specify an expiration time when creating a JWT.

In this book, we'll be using JWTs in two ways. To make it simple, we'll have two types of tokens:

- **An access token**: Used to access resources and handle authorization
- **A refresh token**: Used to retrieve a new access token

But why use two tokens? As we stated earlier, a JWT is generated when users log in. Moreover, JWTs used to access resources should have a short lifespan. This means that after the JWT has expired, the user has to log in again and again – and no user wants the login page to appear every 5 minutes.

That's where a refresh token is useful. It'll contain the essential information needed to verify the user and generate a new access token.

Now that we understand the purpose of JWTs, let's learn more about models in Django while creating the user model.

Organizing a project

When working with Django, you'll have to create many apps to handle different parts of a project. For example, you can have a different application for authentication, and another for payments or articles. To have a clean and well-organized project, we can create a Django application that will contain all the apps we will create for this book.

At the root of the project, run the following command:

```
django-admin startapp core
```

A new application will be created. Remove all the files in this app except for the `apps.py` file and the `__init__.py` file. Inside `apps.py`, add the following line:

core/apps.py

```
from django.apps import AppConfig

class CoreConfig(AppConfig):
    default_auto_field = 'django.db.models.BigAutoField'
    name = 'core'
    label = 'core'
```

Register the apps in the `setting.py` file of the project:

CoreRoot/settings.py

```
# Application definition

INSTALLED_APPS = [
    'django.contrib.admin',
    'django.contrib.auth',
    'django.contrib.contenttypes',
    'django.contrib.sessions',
    'django.contrib.messages',
    'django.contrib.staticfiles',

    'core'
]
```

`INSTALLED_APPS` is a Django settings configuration, which is a list of Django apps within a project.

We can now create the user application with confidence and write our first model.

Creating a user model

Unless you are creating a simple web application, there is little chance of avoiding the necessity to interact with a database, particularly having an account feature that requires users to register or log in to use your web application.

Before talking about the account feature, let's learn more about Django models and what problems they resolve.

What are Django models?

If you need to connect your application to a database, particularly **SQL**, the first assumption that comes to mind is that you'll have to work directly with the database via SQL queries – and if that's true, it can be fun, but it's not the same for everyone; some developers may find SQL complex. You are no longer focusing on writing the application logic in your own language. Some tasks can become repetitive, such as writing SQL scripts to create tables, getting entries from the database, or inserting or updating data.

As you'll see, the more the code base evolves, the more difficult it becomes to maintain both simple and complex SQL queries in your code base. This is more of an issue if you are working with multiple databases, which will require you to learn many SQL languages. For example, there are a lot of SQL databases and each one implements SQL in its own way.

Fortunately, in Django, this messy issue is solved by using a Django model to access the database. This doesn't mean that you don't have to write SQL queries: it's just that you don't have to use SQL at all unless you want to.

Django models provide **object-relational mapping** (**ORM**) to the underlying database. ORM is a tool that simplifies database programming by providing a simple mapping between the object and the database. Then, you don't necessarily need to know the database structure or write complex SQL queries to manipulate or retrieve data from the database.

For example, creating a table in SQL will require writing a long SQL query. Doing this in Python will just require writing a class inheriting from the `django.db` package (*Figure 2.1*):

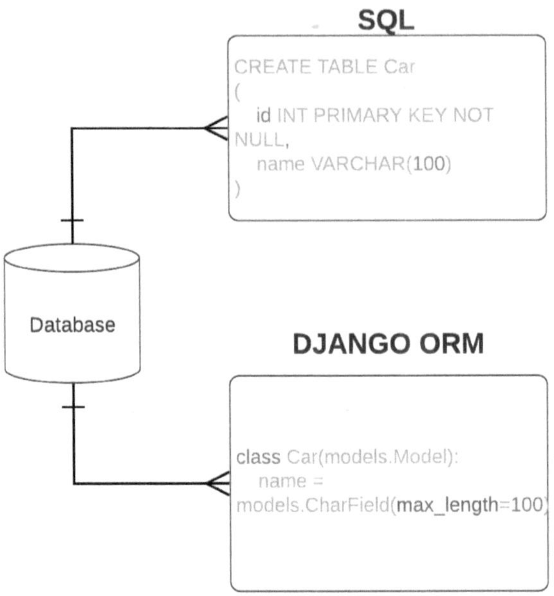

Figure 2.1 – Comparison between the Django ORM and SQL queries

In the preceding figure, you can see the SQL statement, which requires some knowledge of the syntax, as well as the fields and the options. The second code from the Django ORM does exactly the same thing but in a more Pythonic and less verbose manner.

Writing models with Django comes with several advantages:

- **Simplicity**: Writing queries in Python may not be as clear as writing in SQL, but it's less error-prone and more efficient, as you don't have to control which type of database you are working with before trying to understand the code.
- **Consistency**: SQL is inconsistent across different databases. Working with Django models creates an abstraction and helps you focus on the most important tasks.
- **Tracking**: It's even easier to track database design changes working with Django models. It's done by reading migration files written in Python. We'll discuss this more in the next chapter.

Notice that you also have access to model managers. Django Manager is a class that behaves as an interface through which Django models interact with databases. Every Django model, by default, inherits the `models.Manager` class that comes with the necessary methods to make **Create, Read, Update and Delete** (**CRUD**) operations on the table in the database.

Now that we have a better understanding of Django models, let's create the first model in this project, the `User` model. Working with our first model, we'll also learn how to use the basic methods of the Django ORM to perform CRUD operations.

Writing the User model

In the previous section, we saw how a model is represented as a class and how this can basically be created as a table in the database.

Talking about the `User` model, Django comes with a pre-built-in `User` model class that you can use for basic authentication or a session. It actually provides an authentication feature you can use to quickly add authentication and authorization to your projects.

While it's great for most use cases, it has its limitations. For example, in this book, we are building a social media web application. The user in this application will have some bio or even an avatar. Why not also have a phone number for **two-factor authentication** (**2FA**)?

Actually, the `User` model of Django doesn't come with these fields. This means we'll need to extend it and have our own user model. This also means that we will have to add custom methods to the manager for creating a user and a superuser. This will speed up the coding process. In Django, a superuser is a user with administrator permission.

Before creating the model, we actually need an application, and to register it. A Django application is a submodule of a Django project. It's a Python package structured to work in a Django project and share Django conventions such as containing files or submodules such as `models`, `tests`, `urls`, and `views`.

Creating the user application

To start a new application in this project, run the following command:

```
cd core && django-admin startapp user
```

This will create a new package (directory) containing new files. Here's the structure of the directory:

```
├── admin.py
├── apps.py
├── __init__.py
├── migrations
│   └── __init__.py
├── models.py
├── tests.py
└── views.py
```

We can now confidently start writing the `User` model. Here is the structure of the `User` table we want to have in the database:

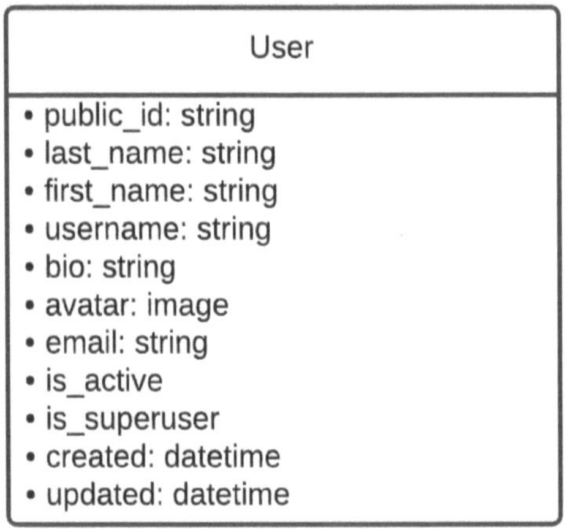

Figure 2.2 – User table structure

And here's the code concerning the `User` table structure:

core/user/models.py

```
import uuid

from django.contrib.auth.models import AbstractBaseUser,
    BaseUserManager, PermissionsMixin
from django.core.exceptions import ObjectDoesNotExist
from django.db import models
from django.http import Http404

class User(AbstractBaseUser, PermissionsMixin):
    public_id = models.UUIDField(db_index=True, unique=True,
        default=uuid.uuid4, editable=False)
    username = models.CharField(db_index=True,
        max_length=255, unique=True)
    first_name = models.CharField(max_length=255)
    last_name = models.CharField(max_length=255)
```

```
    email = models.EmailField(db_index=True, unique=True)
    is_active = models.BooleanField(default=True)
    is_superuser = models.BooleanField(default=False)
    created = models.DateTimeField(auto_now=True)
    updated = models.DateTimeField(auto_now_add=True)

    USERNAME_FIELD = 'email'
    REQUIRED_FIELDS = ['username']

    objects = UserManager()

    def __str__(self):
        return f"{self.email}"

    @property
    def name(self):
        return f"{self.first_name} {self.last_name}"
```

The `models` module from Django provides some field utilities that can be used to write fields and add some rules. For example, `CharField` represents the type of field to create in the `User` table, similar to `BooleanField`. `EmailField` is also `CharField` but rewritten to validate the email that is passed as a value to this field.

We also set the `EMAIL_FIELD` as the email, and `USERNAME_FIELD` as the username. This will help us have two fields for login. The username can be the actual username of a user or just the email address used for registration.

We also have methods such as `name`, which is basically a model property. Then, it can be accessed anywhere on a `User` object, such as `user.name`. We are also rewriting the `__str__` method to return a string that can help us quickly identify a `User` object.

Creating the user and superuser

Next, let's write `UserManager` so we can have methods to create a user and a superuser:

core/user/models.py

```
class UserManager(BaseUserManager):
    def get_object_by_public_id(self, public_id):
```

```
        try:
            instance = self.get(public_id=public_id)
            return instance
        except (ObjectDoesNotExist, ValueError, TypeError):
            return Http404

    def create_user(self, username, email, password=None,
         **kwargs):
        """Create and return a `User` with an email, phone
            number, username and password."""
        if username is None:
            raise TypeError('Users must have a username.')
        if email is None:
            raise TypeError('Users must have an email.')
        if password is None:
            raise TypeError('User must have an email.')

        user = self.model(username=username,
            email=self.normalize_email(email), **kwargs)
        user.set_password(password)
        user.save(using=self._db)

        return user

    def create_superuser(self, username, email, password,
        **kwargs):
        """
        Create and return a `User` with superuser (admin)
            permissions.
        """
        if password is None:
            raise TypeError('Superusers must have a
            password.')
        if email is None:
            raise TypeError('Superusers must have an
```

```
            email.')
    if username is None:
        raise TypeError('Superusers must have an
        username.')

    user = self.create_user(username, email, password,
        **kwargs)
    user.is_superuser = True
    user.is_staff = True
    user.save(using=self._db)

    return user
```

For the `create_user` method, we are basically making sure that fields such as `password`, `email`, `username`, `first_name`, and `last_name` are not `None`. If everything is good, we can confidently call the model, set a password, and save the user in the table.

This is done using the `save()` method.

`create_superuser` also behaves in accordance with the `create_user` method – and it's quite normal because, after all, a superuser is just a user with admin privileges, and also fields such as `is_superuser` and `is_staff` set to `True`. Once it's done, we save the new `User` object in the database and return the user.

See the `save` method as a way to commit changes made to the `User` object to the database.

The model is written and now we need to run migrations to create the table in the database.

Running migrations and testing the model

Before running the migrations, we need to register the user application in `INSTALLED_APPS` in `CoreRoot/settings.py`.

First, let's rewrite the `apps.py` file of the user. It contains the app configs that Django will use to locate the application. Let's also add a label for the application:

core/user/apps.py

```
from django.apps import AppConfig


class UserConfig(AppConfig):
```

```
    default_auto_field = 'django.db.models.BigAutoField'
    name = 'core.user'
    label = 'core_user'
Let's register the application now:
    'core',
    'core.user'
]
```

Let's register the application now in the `INSTALLED_APPS` setting:

CoreRoot/settings.py

```
...
    'core',
    'core.user'
]
```

We also need to tell Django to use this `User` model for the authentication user model. In the `settings.py` file, add the following line:

CoreRoot/settings.py

```
AUTH_USER_MODEL = 'core_user.User'
```

Great – we can now create the first migration for the user app:

```
python manage.py makemigrations
```

You'll have a similar output:

```
Migrations for 'core_user':
  core/user/migrations/0001_initial.py
    - Create model User
```

Let's migrate this modification to the database:

```
python manage.py migrate
```

The table is created in the database. Let's use the Django shell to play with the newly created model a little bit:

```
python manage.py shell
```

Let's import the model and add a dict containing the data needed to create a user:

```
Python 3.10.1 (main, Dec 21 2021, 17:46:38) [GCC 9.3.0] on
linux
Type "help", "copyright", "credits" or "license" for more
information.
(InteractiveConsole)
>>> from core.user.models import User
>>> data_user = {
... "email": "testuser@yopmail.com",
... "username": "john-doe",
... "password": "12345",
... "first_name": "John",
... "last_name": "Doe"
... }
>>> user =  User.objects.create_user(**data_user)


The user is created in the database. Let's access some
properties of the user object.
>>> user.name
'John Doe'
>>> user.email
'testuser@yopmail.com'
>>> user.password
'pbkdf2_sha256$320000$NxM7JZ0cQ0OtDzCVusgvV7$fM1WZp7QhHC3QEajnb
Bjo5rBPKO+Q8ONhDFkCV/gwcI='
```

Great – we've just written the model and created the first user. However, a web browser won't directly read the user data from our database – and worse, we are working with a Python native object, and a browser or a client reaching our server to make requests mostly supports JSON or XML. One idea would be to use the `json` library, but we are dealing with a complex data structure; complex data structures can be easily handled with serializers.

Let's write serializers in the next section.

Writing UserSerializer

A serializer allows us to convert complex Django complex data structures such as `QuerySet` or model instances into Python native objects that can be easily converted to JSON or XML format. However, a serializer also serializes JSON or XML to native Python. **Django Rest Framework** (**DRF**) provides a `serializers` package you can use to write serializers and also validations when API calls are made to an endpoint using this serializer. Let's install the DRF package and make some configurations first:

```
pip install djangorestframework django-filter
```

Don't forget to add the following to the `requirements.txt` file:

requirements.txt

```
Django==4.0.1
psycopg2-binary==2.9.3
djangorestframework==3.13.1
django-filter==21.1
```

We are also adding `django-filter` for data filtering support. Let's add `rest_framework` to the `INSTALLED_APPS` setting:

CoreRoot/settings.py

```
INSTALLED_APPS = [
    ...
    'rest_framework',
]
```

In the `core/user` directory, create a file called `serializers.py`. This file will contain the `UserSerializer` class:

core/user/serializers.py

```
from rest_framework import serializers

from core.user.models import User


class UserSerializer(serializers.ModelSerializer):
```

```
    id = serializers.UUIDField(source='public_id',
        read_only=True, format='hex')
    created = serializers.DateTimeField(read_only=True)
    updated = serializers.DateTimeField(read_only=True)

    class Meta:
        model = User
        fields = ['id', 'username', 'first_name',
            'last_name', 'bio', 'avatar', 'email',
            'is_active', 'created', 'updated']
        read_only_field = ['is_active']
```

The `UserSerializer` class inherits from the `serializers.ModelSerialzer` class. It's a class inheriting from the `serializers.Serializer` class but has deep integrations for supporting a model. It'll automatically match the field of the model to have the correct validations for each one.

For example, we've stated that the email is unique. Then, every time someone registers and enters an email address that already exists in the database, they will receive an error message concerning this.

The `fields` attribute contains all the fields that can be read or written. Then, we also have the `read_only` fields. These fields are only readable. This means that they can't be modified and it's definitely better like that. Why give the external user the possibility to modify the `created`, `updated`, or `id` fields?

Now that `UserSerializer` is available, we can now write `viewset`.

Writing UserViewset

As we know, Django at its core is based on the **Model-View-Template** (**MVT**) architecture. The model communicates with the views (or controllers) and the template displays responses or redirects requests to the views.

However, when Django is coupled with DRF, the model can be directly connected to the view. However, as good practice, use a serializer between a model and a viewset. This really helps with validation and also some important checks.

So, what is a viewset then? DRF provides a class named `APIView` from which a lot of classes from DRF inherit to perform CRUD operations. Therefore, a viewset is simply a class-based view that can handle all the basic HTTP requests—`GET`, `POST`, `PUT`, `DELETE`, and `PATCH`—without hardcoding any CRUD logic here.

For the `viewset` user, we are only allowing the `PATCH` and `GET` methods. Here's what the endpoints will look like:

Method	URL	Result
`GET`	`/api/user/`	Lists all the users
`GET`	`/api/user/user_pk/`	Retrieves a specific user
`PATCH`	`/api/user/user_pk/`	Modifies a user

Table 1.1 – Endpoints

Let's write the viewset. Inside the `user` directory, rename the `view` file `viewsets.py` and add the following content:

core/user/viewsets.py

```
from rest_framework.permissions import AllowAny
from rest_framework import viewsets

from core.user.serializers import UserSerializer
from core.user.models import User


class UserViewSet(viewsets.ModelViewSet):
    http_method_names = ('patch', 'get')
    permission_classes = (AllowAny,)
    serializer_class = UserSerializer

    def get_queryset(self):
        if self.request.user.is_superuser:
            return User.objects.all()
        return User.objects.exclude(is_superuser=True)

    def get_object(self):
     obj =
     User.objects.get_object_by_public_id(self.kwargs['pk'])

        self.check_object_permissions(self.request, obj)

        return obj
```

The only methods allowed here are GET and PUT. We also set serializer_class and permission_classes to AllowAny, which means that anybody can access these viewsets. We also rewrite two methods:

- get_queryset: This method is used by the viewset to get a list of all the users. This method will be called when /user/ is hit with a GET request.
- get_object: This method is used by the viewset to get one user. This method is called when a GET or PUT request is made on the /user/id/ endpoint, with id representing the ID of the user.

There we have the User viewset – but there is no endpoint yet to make it work. Well, let's add a router now.

Adding a router

Routers allow you to quickly declare all of the common routes for a given controller; the next code snippet shows a viewset to which we will be adding a router.

At the root of the apps project (core), create a file named routers.py.

And let's add the code:

core/routers.py

```
from rest_framework import routers
from core.user.viewsets import UserViewSet

router = routers.SimpleRouter()


# ##################################################################
######### #
# ####################
USER                         ###################### #
# ##################################################################
######### #

router.register(r'user', UserViewSet, basename='user')

urlpatterns = [
    *router.urls,
]
```

To register a route for a viewset, the `register()` method needs two arguments:

- **The prefix**: Representing the name of the endpoint, basically
- **The viewset**: Only representing a valid viewset class

The `basename` argument is optional but it's a good practice to use one, as it helps for readability and also helps Django for URL registry purposes.

The router is now added; we can make some requests to the API using Insomnia.

> **Important note**
> Insomnia is a REST client tool used to make requests to RESTful API. With Insomnia, you can manage and create your requests elegantly. It offers support for cookie management, environment variables, code generation, and authentication.

Before doing that, make sure to have the server running:

```
python manage.py runserver
```

Let's make a request to `http://127.0.0.1:8000/api/user/`, a GET request. Look at the following screenshot and make sure to have the same URL – or you can replace `127.0.0.1` with `localhost --`, next to the **Send** button.

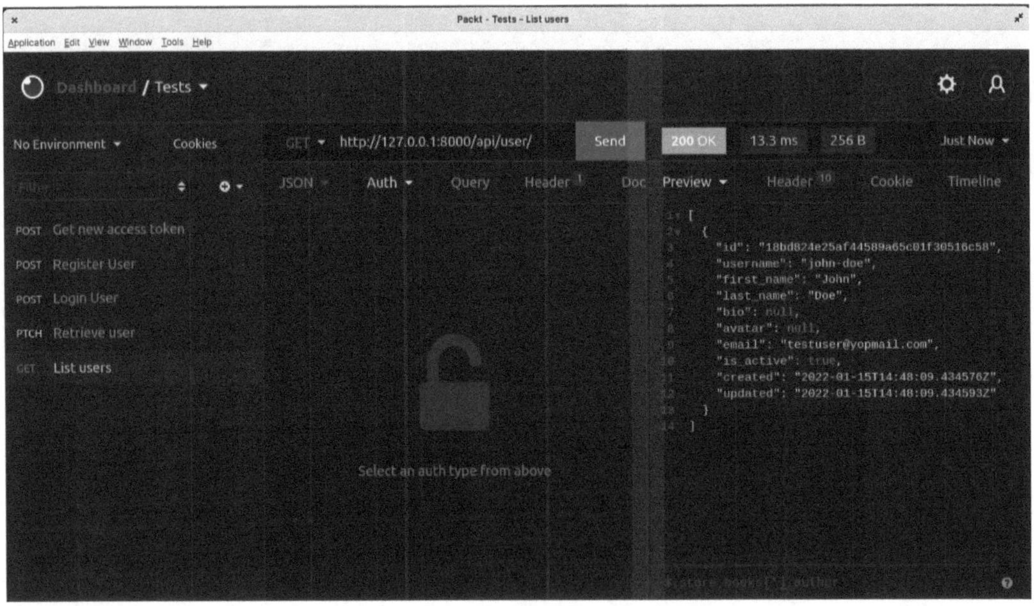

Figure 2.3 – Listing all users

As you can see, we have a list of users created. Let's also make a `GET` request to retrieve the first user using this URL: `/api/user/<id>/`.

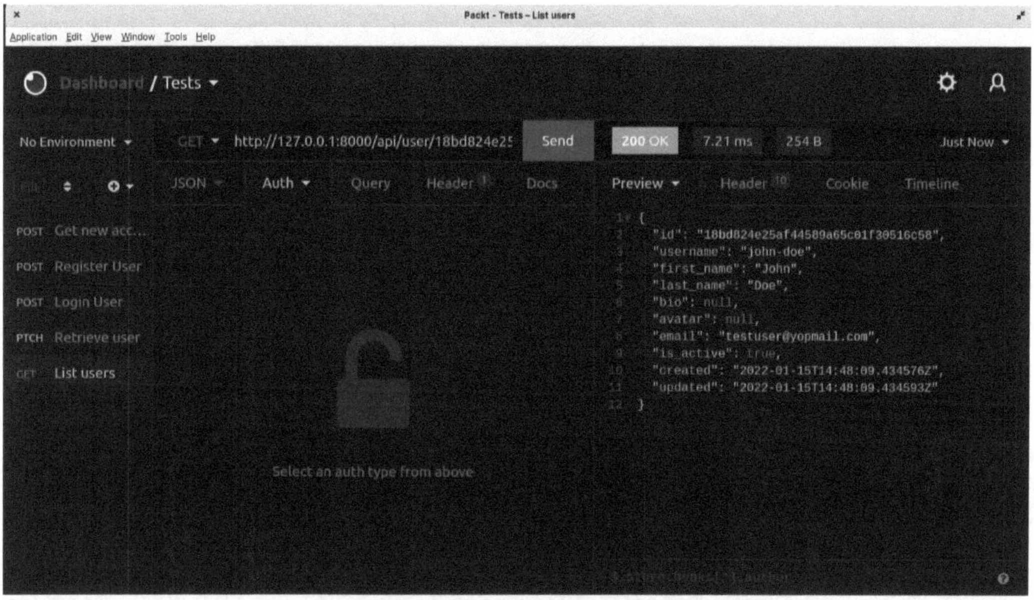

Figure 2.4 – Retrieving a user

We have now a `User` object. This endpoint also allows `PATCH` requests. Let's set the `last_name` value for this user to `Hey`. Change the type of request to `PATCH` and add a JSON body.

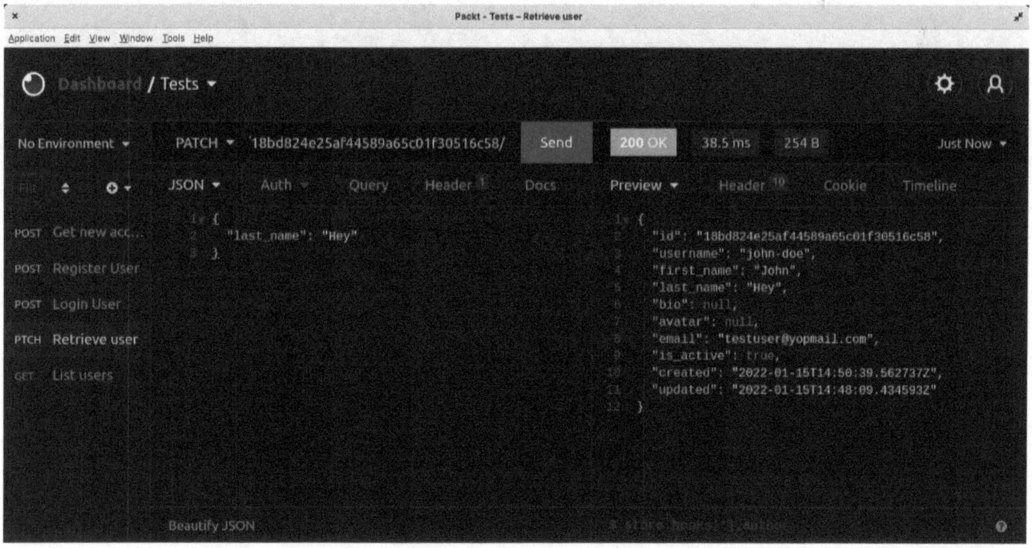

Figure 2.5 – Modifying a user without permissions

Although it's working, it's actually a very bad scenario. We can't have users modify other user names or data. A solution is to change the permission on the `permission_classes` attribute in the `UserViewSet` class:

core/user/viewsets.py

```
from rest_framework.permissions import IsAuthenticated

...
class UserViewSet(viewsets.ModelViewSet):
    http_method_names = ('patch', 'get')
    permission_classes = (IsAuthenticated,)
    serializer_class = UserSerializer
...
```

Let's try the `PATCH` request again.

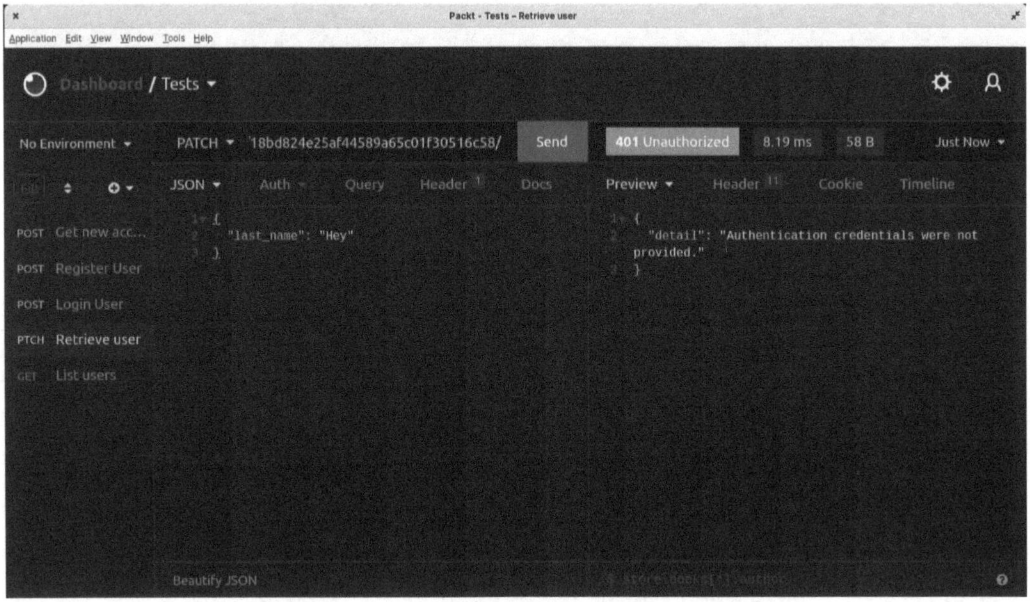

Figure 2.6 – Modifying a user without permissions

We normally have a `401` status, an indication of an authentication issue. Basically, it means that an authentication header should be provided. They are more permissions to add concerning interactions with users, but let's discuss this in later chapters.

Great. Now that we are done with the user application, we can confidently move on to adding a login and registration feature to the project.

Writing the user registration feature

Before accessing protected data, the user needs to be authenticated. This comes with the assumption that there is a registration system to create an account and credentials.

To make things simpler, if the registration of a user is successful, we will provide credentials, here JWTs, so the user won't have to log in again to start a session – a win for user experience.

First, let's install a package that will handle JWT authentication for us. The `djangorestframework-simplejwt` package is a JWT authentication plugin for DRF:

```
pip install djangorestframework-simplejwt
```

The package covers the most common use case of JWT, and in this case here, it facilitates the creation and management of access tokens, as well as refreshing tokens. Before working with this package, there are some configurations needed in the `settings.py` file. We need to register the app in `INSTALLED_APPS` and specify `DEFAULT_AUTHENTICATION_CLASSES` in the `REST_FRAMEWORK` dict:

CoreRoot/settings.py

```
    …
    # external packages apps

    'rest_framework',
    'rest_framework_simplejwt',

    'core',
    'core.user'
]
...
REST_FRAMEWORK = {

    'DEFAULT_AUTHENTICATION_CLASSES': (
        'rest_framework_simplejwt.authentication
            .JWTAuthentication',
    ),
    'DEFAULT_FILTER_BACKENDS':
      ['django_filters.rest_framework.DjangoFilterBackend'],
}
```

First, we need to write a registration serializer, but before that, let's create a new application called `auth` in the `core` app:

```
cd core && django-admin startapp auth
```

It'll contain all the logic concerning logging in, registration, logging out, and a lot more.

As we did earlier for the user application, let's rewrite the `apps.py` file and register the application in the `INSTALLED_APPS` settings:

core/auth/apps.py

```
from django.apps import AppConfig


class AuthConfig(AppConfig):
    default_auto_field = 'django.db.models.BigAutoField'
    name = 'core.auth'
    label = 'core_auth'
And adding the new application to INSTALLED_APPS:
...
'core',
    'core.user',
    'core.auth'
]
...
```

Remove the `admin.py` and `models.py` files from the `auth` directory, as we won't be working with them. For registration and login, we'll have many serializers and viewsets, so let's organize the code accordingly. Create a Python package called `serializers` and another one called `viewsets`. Make sure that these new directories have an `__init__.py` file. Here's how your `auth` app tree should look:

```
├── apps.py
├── __init__.py
├── migrations
│   ├── __init__.py
├── serializers
│   └── __init__.py
├── tests.py
```

```
├── viewsets
│   └── __init__.py
└── views.py
```

Inside the `serializers` directory, create a file called `register.py`. It'll contain the code for `RegisterSerializer`, which is the name of the registration serializer class:

core/auth/serializers/register.py

```
from rest_framework import serializers

from core.user.serializers import UserSerializer
from core.user.models import User

class RegisterSerializer(UserSerializer):
    """
    Registration serializer for requests and user creation
    """

    # Making sure the password is at least 8 characters
        long, and no longer than 128 and can't be read
    # by the user
    password = serializers.CharField(max_length=128,
        min_length=8, write_only=True, required=True)

    class Meta:
        model = User
        # List of all the fields that can be included in a
            request or a response
        fields = ['id', 'bio', 'avatar', 'email',
            'username', 'first_name', 'last_name',
            'password']

    def create(self, validated_data):
        # Use the `create_user` method we wrote earlier for
            the UserManager to create a new user.
        return User.objects.create_user(**validated_data)
```

As you can see, RegisterSerializer is a subclass of UserSerializer. This is really helpful, as we don't need to rewrite fields again.

Here, we don't need to revalidate fields such as email or password. As we declared these fields with some conditions, Django will automatically handle their validation.

Next, we can add the viewset and register it in the register.py file:

core/auth/viewsets/register.py

```
from rest_framework.response import Response
from rest_framework.viewsets import ViewSet
from rest_framework.permissions import AllowAny
from rest_framework import status
from rest_framework_simplejwt.tokens import RefreshToken
from core.auth.serializers import RegisterSerializer


class RegisterViewSet(ViewSet):
    serializer_class = RegisterSerializer
    permission_classes = (AllowAny,)
    http_method_names = ['post']

    def create(self, request, *args, **kwargs):
        serializer =
            self.serializer_class(data=request.data)

        serializer.is_valid(raise_exception=True)
        user = serializer.save()
        refresh = RefreshToken.for_user(user)
        res = {
            "refresh": str(refresh),
            "access": str(refresh.access_token),
        }

        return Response({
            "user": serializer.data,
```

```
            "refresh": res["refresh"],
            "token": res["access"]
        }, status=status.HTTP_201_CREATED)
```

Nothing really new here – we are using attributes from the `ViewSet` class. We are also rewriting the `create` method to add access and refresh tokens in the body of the response. The `djangorestframework-simplejwt` package provides utilities we can use to directly generate tokens. That's what `RefreshToken.for_user(user)` does.

And the final step – let's register the viewset in the `routers.py` file:

core/routers.py

```
...
# ##################################################################
######### #
# ####################
AUTH                          ######################## #
# ##################################################################
######### #

router.register(r'auth/register', RegisterViewSet,
    basename='auth-register')
...
```

Great! Let's test the new endpoint with Insomnia. In the collection of requests for this project, create a new `POST` request. The URL will be as follows: `localhost:8000/api/auth/register/`.

As a body for the request, you can pass the following:

```
{
    "username": "mouse21",
    "first_name": "Mickey",
    "last_name": "Mouse",
    "password": "12345678",
    "email": "mouse@yopmail.com"
}
```

With that, send the request. You should have a response similar to that shown in *Figure 2.6* with a `201 HTTP` status:

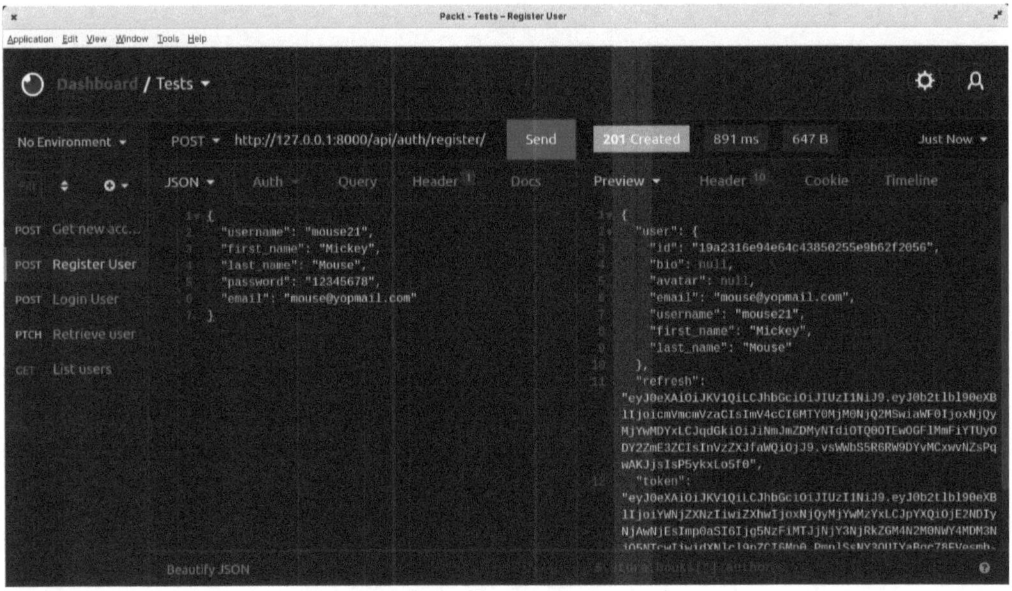

Figure 2.7 – Registering a user

Let's see what happens if we try to create a user with the same email and username. Hit the **Send** button to send the same request again. You should receive a `400` error.

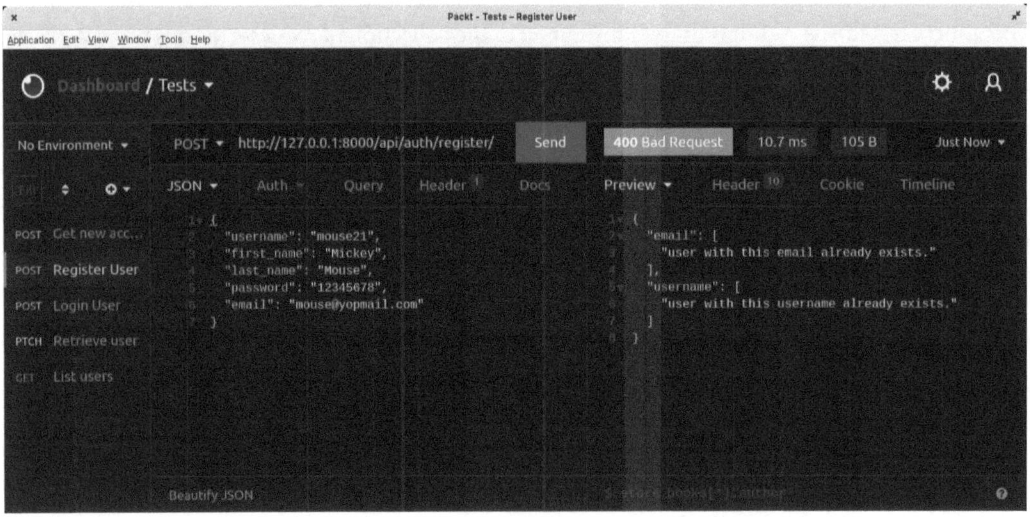

Figure 2.8 – Registering a user with the same email and username

Great. We are now sure that the endpoint behaves as we wish. The next step will be to add the login endpoint following the same process: writing the serializer and the viewset, and then registering the route.

Adding the login feature

The login feature will require the email or the username with the password. Using the `djangorestframework-simplejwt` package, which provides a serializer called `TokenObtainPairSerializer`, we'll write a serializer to check for user authentication but also return a response containing access and refresh tokens. For this, we will rewrite the validate method from the `TokenObtainPairSerializer` class. Inside the `core/auth/serializers` directory, create a new file called `login.py` (this file will contain `LoginSerializer`, a subclass of `TokenObtainPairSerializer`):

core/auth/serializers/login.py

```
from rest_framework_simplejwt.serializers import
  TokenObtainPairSerializer
from rest_framework_simplejwt.settings import api_settings
from django.contrib.auth.models import update_last_login

from core.user.serializers import UserSerializer


class LoginSerializer(TokenObtainPairSerializer):

    def validate(self, attrs):
        data = super().validate(attrs)

        refresh = self.get_token(self.user)

        data['user'] = UserSerializer(self.user).data
        data['refresh'] = str(refresh)
        data['access'] = str(refresh.access_token)

        if api_settings.UPDATE_LAST_LOGIN:
            update_last_login(None, self.user)

        return data
```

We are surcharging the `validate` method from the `TokenObtainPairSerializer` class to adapt it to our needs. That's why `super` is helpful here. It's a built-in method in Python that returns a temporary object that can be used to access the class methods of the base class.

Then, we use `user` to retrieve access and refresh tokens. Once the serializer is written, don't forget to import it to the `__init__.py` file:

core/auth/serializers/__init__.py

```
from .register import RegisterSerializer
from .login import LoginSerializer
```

The next step is to add the viewset. We'll call this viewset `LoginViewset`. As we are not directly interacting with a model here, we'll just be using the `viewsets.ViewSet` class:

core/auth/viewsets/login.py

```
from rest_framework.response import Response
from rest_framework.viewsets import ViewSet
from rest_framework.permissions import AllowAny
from rest_framework import status
from rest_framework_simplejwt.exceptions import TokenError,
    InvalidToken
from core.auth.serializers import LoginSerializer


class LoginViewSet(ViewSet):
   serializer_class = LoginSerializer
   permission_classes = (AllowAny,)
   http_method_names = ['post']

   def create(self, request, *args, **kwargs):
       serializer =
           self.serializer_class(data=request.data)

       try:
           serializer.is_valid(raise_exception=True)
       except TokenError as e:
           raise InvalidToken(e.args[0])
```

```
        return Response(serializer.validated_data,
            status=status.HTTP_200_OK)
```

Add the viewset to the `__init__.py` file of the `viewsets` directory:

```
from .register import RegisterViewSet
from .login import LoginViewSet
```

We can now import it and register it in the `routers.py` file:

core/routers.py

```
...
from core.auth.viewsets import RegisterViewSet,
    LoginViewSet

router = routers.SimpleRouter()

# ##################################################################
######### #
# ####################
AUTH                       ####################### #
# ##################################################################
######### #

router.register(r'auth/register', RegisterViewSet,
    basename='auth-register')
router.register(r'auth/login', LoginViewSet,
    basename='auth-login')
...
```

The endpoint for login will be available at `/auth/login/`. Let's try a request with Insomnia.

Here's the body of the request I'll use:

```
{
    "password": "12345678",
    "email": "mouse@yopmail.com"
}
```

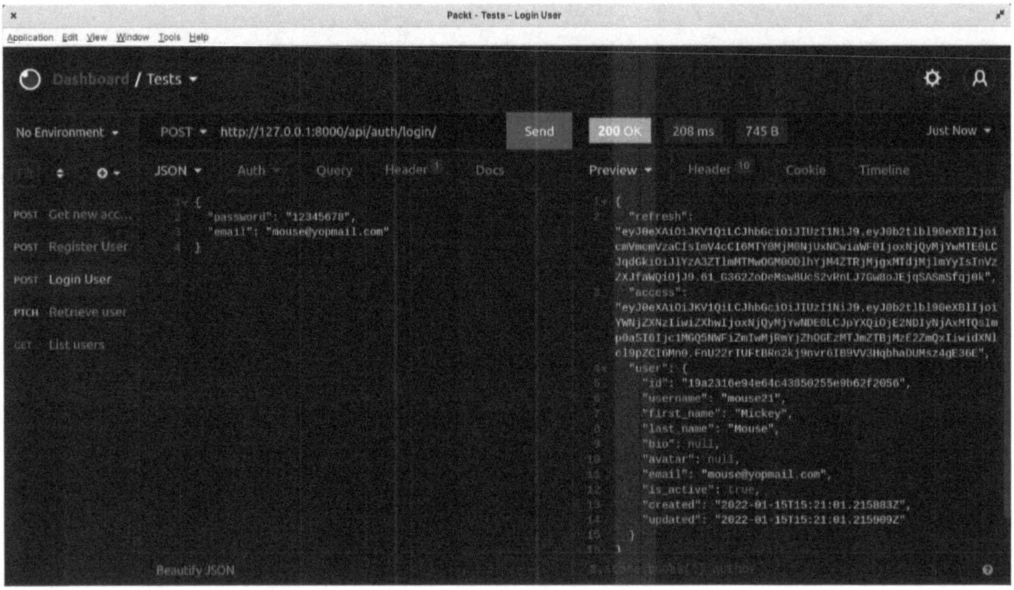

Figure 2.9 – Login with user credentials

The login feature is ready and working like a charm – but we have a little bit of an issue. The access token expires in 5 minutes. Basically, to get a new access token, the user will have to log in again. Let's see how we can use the refresh token to request a new access token without logging in again.

Refresh logic

`djangorestframework-simplejwt` provides a refresh logic feature. As you've noticed, we've been generating refresh tokens and returning them as responses every time registration or login is completed. We'll just inherit the class from `TokenRefreshView` and transform it into a viewset.

In `auth/viewsets`, add a new file called `refresh.py`:

core/auth/viewsets/refresh.py

```
from rest_framework.response import Response
from rest_framework_simplejwt.views import TokenRefreshView
from rest_framework.permissions import AllowAny
from rest_framework import status
from rest_framework import viewsets
from rest_framework_simplejwt.exceptions import TokenError,
    InvalidToken
```

```
class RefreshViewSet(viewsets.ViewSet, TokenRefreshView):
    permission_classes = (AllowAny,)
    http_method_names = ['post']

    def create(self, request, *args, **kwargs):
        serializer = self.get_serializer(data=request.data)

        try:
            serializer.is_valid(raise_exception=True)
        except TokenError as e:
            raise InvalidToken(e.args[0])

        return Response(serializer.validated_data,
            status=status.HTTP_200_OK)
Now add the class in the __init__.py file.
from .register import RegisterViewSet
from .login import LoginViewSet
from .refresh import RefreshViewSet
```

Now add the class in the `__init__.py` file.

core/auth/viewsets/__init__.py

```
from .register import RegisterViewSet
from .login import LoginViewSet
from .refresh import RefreshViewSet
```

And now register it in the `routers.py` file:

core/routers.py

```
from core.auth.viewsets import RegisterViewSet,
    LoginViewSet, RefreshViewSet
...
router.register(r'auth/refresh', RefreshViewSet,
    basename='auth-refresh')
...
```

Great – let's test the new endpoint at `/auth/refresh/` to get a new token. It'll be a `POST` request with the refresh token in the body of the request, and you will receive a new access token in the response:

Figure 2.10 – Requesting for a new access token

Great – we've just learned how to implement refresh token logic in the application.

Summary

In this chapter, we learned how to write an authentication system based on JWT for a Django application using DRF and `djangorestframework-simplejwt`. We also learned how to extend classes and rewrite the functions.

In the next chapter, we'll add the `posts` feature. Our users will be able to create a post that can be viewed and liked by other users.

Questions

1. What is a JWT?
2. What is Django Rest Framework?
3. What is a model?
4. What is a serializer?
5. What is a viewset?
6. What is a router?
7. What is the usage of a refresh token?

3

Social Media Post Management

In the previous chapter, we introduced models, serializers, viewsets, and routes to create our first endpoints. In this chapter, we will be working with the same concepts for creating posts for our social media project. This will be done by dividing the project into concepts such as database relations, filtering, and permissions. By the end of this chapter, you'll be able to work with database relations with Django models, write custom filters and permissions, and delete and update objects.

We will be covering the following topics in this chapter:

- Creating the Post model
- Writing the Post model
- Writing the Post serializer
- Writing Post viewsets
- Adding permissions
- Deleting and updating posts
- Adding the Like feature

Technical requirements

For this chapter, you need to have Insomnia installed on your machine to make HTTP requests.

You can find the code for this chapter here: `https://github.com/PacktPublishing/Full-stack-Django-and-React/tree/chap3`.

Creating the Post model

A post in this project is a long or short piece of text that can be viewed by anyone, irrespective of whether a user is linked or associated to that post. Here are the requirements for the post feature:

- Authenticated users should be able to create a post
- Authenticated users should be able to like the post
- All users should be able to read the post, even if they aren't authenticated
- The author of the post should be able to modify the post
- The author of the post should be able to delete the post

Looking at these requirements from a backend perspective, we can understand that we'll be dealing with a database, a model, and permissions. First, let's start by writing the structure of the `Post` model in the database.

Designing the Post model

A post consists of content made up of characters written by an author (here, a user). How does that schematize itself into our database?

Before creating the `Post` model, let's draw a quick figure of the structure of the model in the database:

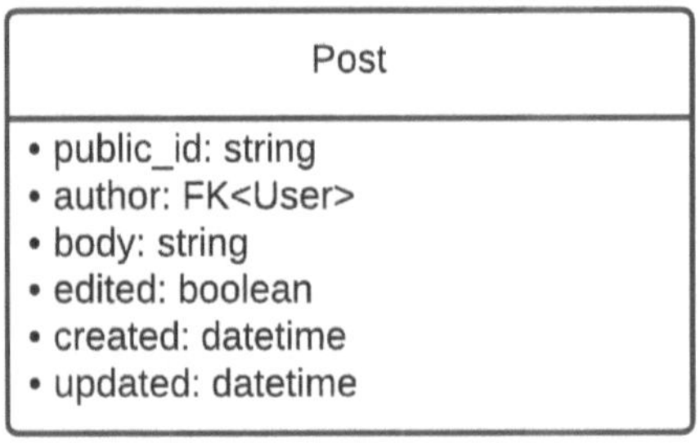

Figure 3.1 – Post table

As you can see in *Figure 3.1*, there is an `author` field, which is a **foreign key**. A foreign key is a set of attributes in a table that refers to the primary key of another table. In our case, the foreign key will refer to the primary key of the `User` table. Each time a post is created, a foreign key will need to be passed.

The foreign key is one of the characteristics of the **one-to-many** (or **many-to-one**) relationship. In this relationship, a row in table A can have many matching rows in table B (*one-to-many*) but a row in table B can only have one matching row in table A.

In our case, a user (from the `User` table) can have many posts (in the `Post` table) but a post can only have one user (*Figure 3.2*):

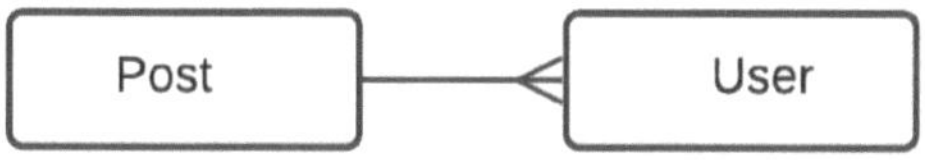

Figure 3.2 – User and Post relationship

There are also two other types of database relationships:

- One-to-one: In this type of relationship, a row in table A can only have one matching row in table B, and vice versa. An example of this can be worker C having one and only one desk D. And this desk D can only be used by this worker C (*Figure 3.3*):

Figure 3.3 – One-to-one relationship between a worker and a desk

- Many-to-many: In this type of database relationship, a row in table A can have many matching rows in table B, and vice versa. For example, in an e-commerce application, an order can have many items, and an item can also appear in many different orders (*Figure 3.4*):

Figure 3.4 – Many-to-many relationship between an order and an item

The *many-to-many* relationship will be used when writing the *like* feature for the posts.

Great, now that we have a better idea of database relationships, we can begin to write the post feature, starting from the `Post` model. But before that, let's quickly refactor the code to make development easier.

Abstraction

The next models that we'll create will also have the `public_id`, `created`, and `updated` fields. For the sake of the **don't repeat yourself** (**DRY**) principle, we will use abstract model classes.

An **abstract class** can be considered a blueprint for other classes. It usually contains a set of methods or attributes that must be created within any child classes built from the abstract class.

Inside the `core` directory, create a new Python package called `abstract`. Once it's done, create a `models.py` file. In this file, we will write two classes: `AbstractModel` and `AbstractManager`.

The `AbstractModel` class will contain fields such as `public_id`, `created`, and `updated`. On the other side, the `AbstractManager` class will contain the function used to retrieve an object by its `public_id` field:

core/abstract/models.py

```
from django.db import models
import uuid

from django.core.exceptions import ObjectDoesNotExist
from django.http import Http404


class AbstractManager(models.Manager):
    def get_object_by_public_id(self, public_id):
        try:
            instance = self.get(public_id=public_id)
            return instance
        except (ObjectDoesNotExist, ValueError, TypeError):
            return Http404


class AbstractModel(models.Model):
    public_id = models.UUIDField(db_index=True, unique=True,
      default=uuid.uuid4, editable=False)
    created = models.DateTimeField(auto_now_add=True)
    updated = models.DateTimeField(auto_now=True)


    objects = AbstractManager()

    class Meta:
        abstract = True
```

As you can see in the `Meta` class for `AbstractModel`, the `abstract` attribute is set to `True`. Django will ignore this class model and won't generate migrations for this.

Now that we have this class, let's make a quick refactor on the User model:

First, let's remove the get_object_by_public_id method to retrieve an object via public_id, and let's subclass UserManager:

core/user/models.py

```
...
from core.abstract.models import AbstractModel, AbstractManager


class UserManager(BaseUserManager, AbstractManager):
...
class User(AbstractModel, AbstractBaseUser, PermissionsMixin):
...
```

On the User model, remove the public_id, updated, and created fields, and also, subclass the User model with the AbstractModel class. This will normally cause no changes to the database, hence, there is no need to run makemigrations again unless you've changed an attribute of a field.

Let's also add AbstractSerializer, which will be used by all the serializers we'll be creating on this project.

Writing the AbstractSerializer

All the objects sent back as a response in our API will contain the id, created, and updated fields. It'll be repetitive to write these fields all over again on every ModelSerializer, so let's just create an AbstractSerializer class. In the abstract directory, create a file called serializers.py and add the following content:

core/abstract/serializers.py

```
from rest_framework import serializers

class AbstractSerializer(serializers.ModelSerializer):
    id = serializers.UUIDField(source='public_id',
                               read_only=True, format='hex')
    created = serializers.DateTimeField(read_only=True)
    updated = serializers.DateTimeField(read_only=True)
```

Once it's done, you can go and subclass the `UserSerializer` class with the `AbstractSerializer` class:

core/user/serializers.py

```
from core.abstract.serializers import AbstractSerializer
from core.user.models import User

class UserSerializer(AbstractSerializer):
...
```

Once it's done, remove the field declaration of `id`, `created`, and `updated`.

Let's perform one last abstraction for `ViewSets`.

Writing the AbstractViewSet

But why write an abstract `ViewSet`? Well, there will be repeated declarations as to the ordering and the filtering. Let's create a class that will contain the default values.

In the `abstract` directory, create a file called `viewsets.py` and add the following content:

core/abstract/viewsets.py

```
from rest_framework import viewsets
from rest_framework import filters

class AbstractViewSet(viewsets.ModelViewSet):
    filter_backends = [filters.OrderingFilter]
    ordering_fields = ['updated', 'created']
    ordering = ['-updated']
```

As you can see, we have the following attributes:

- `filter_backends`: This sets the default filter backend.
- `ordering_fields`: This list contains the fields that can be used as ordering parameters when making a request.
- `ordering`: This will tell Django REST in which order to send many objects as a response. In this case, all the responses will be ordered by the most recently updated.

The next step is to add the `AbstractViewSet` class to the code where `ModelViewSets` is actually called. Go to `core/user/viewsets.py` and subclass `UserViewSet` with the `AbstractViewSet` class:

core/user/viewsets.py

```
...
from core.abstract.viewsets import AbstractViewSet
from core.user.serializers import UserSerializer
from core.user.models import User

class UserViewSet(AbstractViewSet):
...
```

Great, now we have all the things needed to write better and less code; let's write the `Post` model.

Writing the Post model

We have already established the structure of the `Post` model. Let's write the code and the features:

1. Create a new application called `post`:

```
django-admin startapp post
```

2. Rewrite `apps.py` of the new create package so it can be called easily in the project:

core/post/apps.py

```
from django.apps import AppConfig

class PostConfig(AppConfig):
    default_auto_field =
      'django.db.models.BigAutoField'
    name = 'core.post'
    label = "core_label"
```

3. Once it's done, we can now write the `Post` model. Open the `models.py` file and enter the following content:

core/post/models.py

```
from django.db import models

from core.abstract.models import AbstractModel,
```

```
AbstractManager

class PostManager(AbstractManager):
    pass

class Post(AbstractModel):
    author = models.ForeignKey(to="core_user.User",
      on_delete=models.CASCADE)
    body = models.TextField()
    edited = models.BooleanField(default=False)

    objects = PostManager()
    def __str__(self):
        return f"{self.author.name}"

    class Meta:
        db_table = "'core.post'"
```

You can see here how we created the `ForeignKey` relationship. Django models actually provide tools to handle this kind of relationship, and it's also symmetrical, meaning that not only can we use the `Post.author` syntax to access the user object but we can also access posts created by a user using the `User.post_set` syntax. The latter syntax will return a `queryset` object containing the posts created by the user because we are in a `ForeignKey` relationship, which is also a one-to-many relationship. You will also notice the `on_delete` attribute with the `models.CASCADE` value. Using `CASCADE`, if a user is deleted from the database, Django will also delete all records of posts in relation to this user.

Apart from `CASCADE` as a value for the `on_delete` attribute on a `ForeignKey` relationship, you can also have the following:

- `SET_NULL`: This will set the child object foreign key to null on delete. For example, if a user is deleted from the database, the value of the `author` field of the posts in relation to this user is set to **None**.
- `SET_DEFAULT`: This will set the child object to the default value given while writing the model. It works if you are sure that the default value won't be deleted.
- `RESTRICT`: This raises `RestrictedError` under certain conditions.
- `PROTECT`: This prevents the foreign key object from being deleted as long as there are objects linked to the foreign key object.

Let's test the newly added model by creating an object and saving it in the database:

1. Add the newly created application to the `INSTALLED_APPS` list:

CoreRoot/settings.py

```
...
'core.post'
...
```

2. Let's create the migrations for the newly added application:

```
python manage makemigrations && python manage.py migrate
```

3. Then, let's play with the **Django shell** by starting it with the `python manage.py shell` command:

```
(venv) koladev@koladev123xxx:~/PycharmProjects/Full-stack-Django-and-React$ python manage.py shell
Python 3.10.2 (main, Jan 15 2022, 18:02:07) [GCC 9.3.0] on linux
Type "help", "copyright", "credits" or "license" for more information.
(InteractiveConsole)
>>>
```

> **Important note**
>
> You can use the **django_shell_plus** package to speed up work with Django shell. You won't need to type all imports yourself as all your models will be imported by default. You can find more information on how to install it from the following website: `https://django-extensions.readthedocs.io/en/latest/shell_plus.html`.

4. Let's import a user. This will be the author of the post we'll be creating:

```
>>> from core.post.models import Post
>>> from core.user.models import User
>>> user = User.objects.first()
>>> user
```

5. Next, let's create a dictionary that will contain all the fields needed to create a post:

```
>>> data = {"author": user, "body":"A simple test"}
```

6. And now, let's create a post:

```
>>> post = Post.objects.create(**data)
>>> post
<Post: John Hey>
>>>
Let's access the author field of this object.
>>> post.author
<User: testuser@yopmail.com>
```

As you can see, the author is in fact the user we've retrieved from the database.

Let's also try the inverse relationship:

```
>>> user.post_set.all()
<QuerySet [<Post: John Hey>]>
```

As you can see, the `post_set` attribute contains all the instructions needed to interact with all the posts linked to this user.

Now that you have a better understanding of how database relationships work in Django, we can move on to writing the serializer of the `Post` object.

Writing the Post serializer

The `Post` serializer will contain the fields needed to create a post when making a request on the endpoint. Let's add the feature for the post creation first.

In the `post` directory, create a file called `serializers.py`. Inside this file, add the following content:

core/post/serializers.py

```
from rest_framework import serializers
from rest_framework.exceptions import ValidationError

from core.abstract.serializers import AbstractSerializer
from core.post.models import Post
from core.user.models import User

class PostSerializer(AbstractSerializer):
    author = serializers.SlugRelatedField(
      queryset=User.objects.all(), slug_field='public_id')
```

```
    def validate_author(self, value):
        if self.context["request"].user != value:
            raise ValidationError("You can't create a post
                                   for another user.")
        return value
    class Meta:
        model = Post
        # List of all the fields that can be included in a
        # request or a response
        fields = ['id', 'author', 'body', 'edited',
                  'created', 'updated']
        read_only_fields = ["edited"]
```

We've added a new serializer field type, `SlugRelatedField`. As we are working with the `ModelSerializer` class, Django automatically handles the fields and relationship generation for us. Defining the type of relationship field we want to use can also be crucial to tell Django exactly what to do.

And that's where `SlugRelatedField` comes in. It is used to represent the target of the relationship using a field on the target. Thus, when creating a post, `public_id` of the author will be passed in the body of the request so that the user can be identified and linked to the post.

The `validate_author` method checks validation for the `author` field. Here, we want to make sure that the user creating the post is the same user as in the `author` field. A context dictionary is available in every serializer. It usually contains the request object that we can use to make some checks.

There is no hard limitation here so we can easily move to the next part of this feature: writing the `Post` viewsets.

Writing Post viewsets

For the following endpoint, we'll only be allowing the `POST` and `GET` methods. This will help us have the basic features working first.

The code should follow these rules:

- Only authenticated users can create posts
- Only authenticated users can read posts
- Only `GET` and `POST` methods are allowed

Inside the `post` directory, create a file called `viewsets.py`. Into the file, add the following content:

core/post/viewsets.py

```
from rest_framework.permissions import IsAuthenticated

from core.abstract.viewsets import AbstractViewSet
from core.post.models import Post
from core.post.serializers import PostSerializer

class PostViewSet(AbstractViewSet):
    http_method_names = ('post', 'get')
    permission_classes = (IsAuthenticated,)
    serializer_class = PostSerializer

    def get_queryset(self):
        return Post.objects.all()

    def get_object(self):
        obj = Post.objects.get_object_by_public_id(
          self.kwargs['pk'])

        self.check_object_permissions(self.request, obj)

        return obj

    def create(self, request, *args, **kwargs):
        serializer = self.get_serializer(data=request.data)
        serializer.is_valid(raise_exception=True)
        self.perform_create(serializer)
        return Response(serializer.data,
                        status=status.HTTP_201_CREATED)
```

In the preceding code, we defined three interesting methods:

- The `get_queryset` method returns all the posts. We don't actually have particular requirements for fetching posts, so we can return all posts in the database.
- The `get_object` method returns a `post` object using `public_id` that will be present in the URL. We retrieve this parameter from the `self.kwargs` directory.
- The `create` method, which is the `ViewSet` action executed on `POST` requests on the endpoint linked to `ViewSet`. We simply pass the data to the serializer declared on `ViewSet`, validate the data, and then call the `perform_create` method to create a `post` object. This method will automatically handle the creation of a `post` object by calling the `Serializer.create` method, which will trigger the creation of a `post` object in the database. Finally, we return a response with the newly created post.

And right here, you have the code for `ViewSet`. The next step is to add an endpoint and start testing the API.

Adding the Post route

In the `routers.py` file, add the following content:

core/routers.py

```
...
from core.post.viewsets import PostViewSet


# ############################################################
######### #
# ###################
POST                       ###################### #
# ############################################################
######### #


router.register(r'post', PostViewSet, basename='post')
...
```

Once it's done, you'll have a new endpoint available on `/post/`. Let's play with Insomnia to test the API.

First of all, try to make a request directly to the `/post/` endpoint. You'll receive a **401 error**, meaning that you must provide an access token. No problem, log in on the `/auth/login/` endpoint with a registered user and copy the token.

In the **Bearer** tab in Insomnia, select **Bearer Token**:

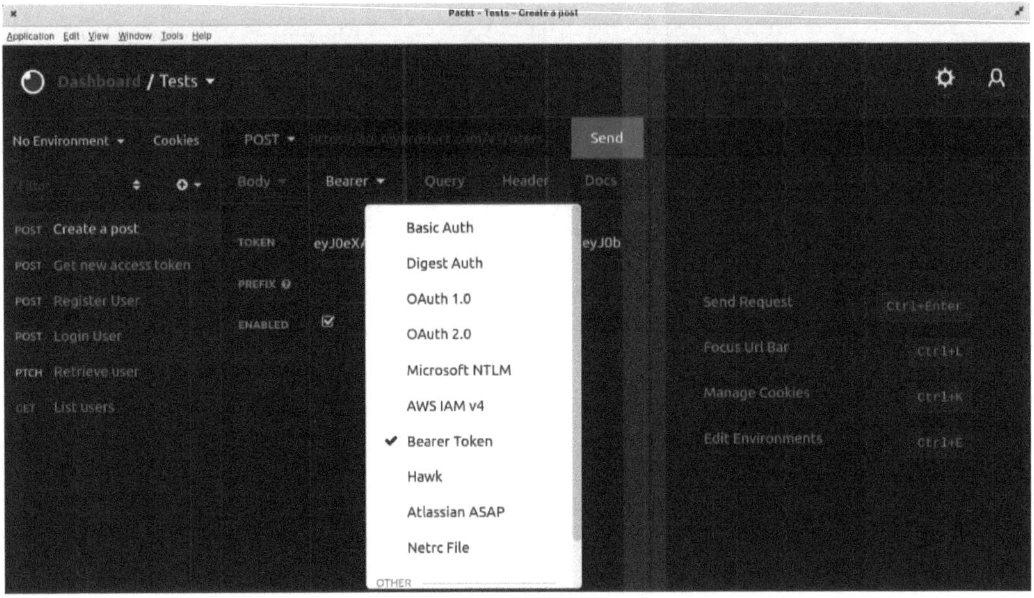

Figure 3.5 – Adding Bearer Token to Insomnia request

Now, fire the endpoint again with a `GET` request. You'll see no results, great! Let's create the first post in the database.

Change the type of request to `POST` and the following to the JSON body:

```
{
    "author": "19a2316e94e64c43850255e9b62f2056",
    "body": "A simple posted"
}
```

Please note that we will have a different `public_id` so make sure to use `public_id` of the user you've just logged in as and send the request again:

Figure 3.6 – Creating a post

Great, the post is created! Let's see whether it's available when making a GET request:

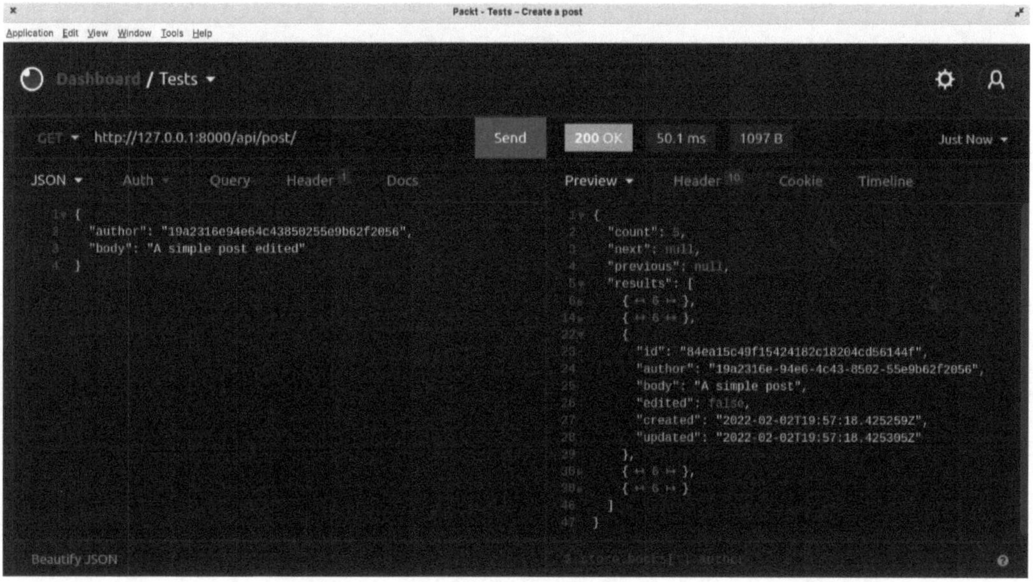

Figure 3.7 – Getting all posts

The DRF provides a way to paginate responses and a default pagination limit size globally in the `settings.py` file. With time, a lot of objects will be shown and the size of the payload will vary.

To prevent this, let's add a default size and a class to paginate our results.

Inside the `settings.py` file of the project, add new settings to the `REST_FRAMEWORK` dictionary:

CoreRoot/settings.py

```
REST_FRAMEWORK = {
...
    'DEFAULT_PAGINATION_CLASS':
      'rest_framework.pagination.LimitOffsetPagination',
    'PAGE_SIZE': 15,
}
...
```

Basically here, all results are limited to 15 per page but we can also increase this size with the `limit` parameter when making a request and also use the `offset` parameter to precisely where we want the result to start from:

```
GET https://api.example.org/accounts/?limit=100&offset=400
```

Great, now make a `GET` request again and you'll see that the results are better structured.

Also, it'll be more practical to have the name of the author in the response as well. Let's rewrite a serializer method that can help modify the response object.

Rewriting the Post serialized object

Actually, the `author` field accepts `public_id` and returns `public_id`. While it does the work, it can be a little bit difficult to identify the user. This will cause it to make a request again with `public_id` of the user to get the pieces of information about the user.

The `to_representation()` method takes the object instance that requires serialization and returns a primitive representation. This usually means returning a structure of built-in Python data types. The exact types that can be handled depend on the render classes you configure for your API.

Inside `post/serializers.py`, add a new method called `to_represenation()`:

core/post/serializers.py

```
class PostSerializer(AbstractSerializer):
    ...

    def to_representation(self, instance):
        rep = super().to_representation(instance)
```

```
        author = User.objects.get_object_by_public_id(
          rep["author"])
        rep["author"] = UserSerializer(author).data

        return rep
...
```

As you can see, we are using the `public_id` field to retrieve the user and then serialize the `User` object with `UserSerializer`.

Let's get all the posts again and you'll see all the users:

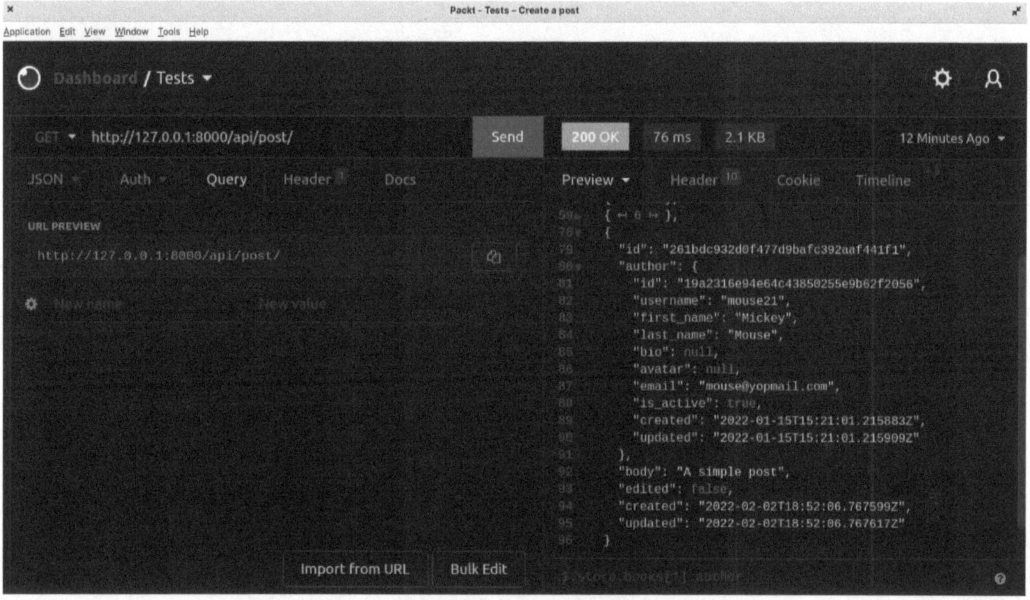

Figure 3.8 – Getting all posts

We have a working `Post` feature but it also has some issues. Let's explore this further when writing permissions for our feature.

Adding permissions

If authentication is the action of verifying the identity of a user, authorization is simply the action of checking whether the user has the rights or privileges to perform an action.

In our project, we have three types of users:

- **The anonymous user**: This user has no account on the API and can't really be identified
- **The registered and active user**: This user has an account on the API and can easily perform some actions
- **The admin user**: This user has all rights and privileges

We want anonymous users to be able to read the posts on the API without necessarily being authenticated. While it's true that there is the `AllowAny` permission, it'll surely conflict with the `IsAuthenticated` permission.

Thus, we need to write a custom permission.

Inside the `authentication` directory, create a file called `permissions`, and add the following content:

core/post/viewsets.py

```
from rest_framework.permissions import BasePermission, SAFE_
METHODS

class UserPermission(BasePermission):
    def has_object_permission(self, request, view, obj):
        if request.user.is_anonymous:
            return request.method in SAFE_METHODS

        if view.basename in ["post"]:
            return bool(request.user and
                        request.user.is_authenticated)
     return False

    def has_permission(self, request, view):
        if view.basename in ["post"]:
            if request.user.is_anonymous:
                return request.method in SAFE_METHODS

            return bool(request.user and
                        request.user.is_authenticated)

        return False
```

Django permissions usually work on two levels: on the overall endpoint (`has_permission`) and on an object level (`has_object_permission`).

A great way to write permissions is to always deny by default; that is why we always return `False` at the end of each permission method. And then you can start adding the conditions. Here, in all the methods, we are checking that anonymous users can only make the `SAFE_METHODS` requests — `GET`, `OPTIONS`, and `HEAD`.

And for other users, we are making sure that they are always authenticated before continuing. Another important feature is to allow users to delete or update posts. Let's see how we can add this with Django.

Deleting and updating posts

Deleting and updating articles are also part of the features of posts. To add these functionalities, we don't need to write a serializer or a viewset, as the methods for deletion (`destroy()`), and updating (`update()`) are already available by default in the `ViewSet` class. We will just rewrite the `update` method on `PostSerializer` to ensure that the `edited` field is set to `True` when modifying a post.

Let's add the `PUT` and `DELETE` methods to `http_methods` of `PostViewSet`:

core/post/viewsets.py

```
...
class PostViewSet(AbstractViewSet):
    http_method_names = ('post', 'get', 'put', 'delete')
...
```

Before going in, let's rewrite the `update` method in `PostSerializer`. We actually have a field called `edited` in the `Post` model. This field will tell us whether the post has been edited:

core/post/serializers.py

```
...
class PostSerializer(AbstractSerializer):
...
    def update(self, instance, validated_data):
        if not instance.edited:
            validated_data['edited'] = True

        instance = super().update(instance, validated_data)
```

```
        return instance
...
```

And let's try the PUT and DELETE requests in Insomnia. Here's an example of the body for the PUT request:

```
{
    "author": "61c5a1ecb9f5439b810224d2af148a23",
    "body": "A simple post edited"
}
```

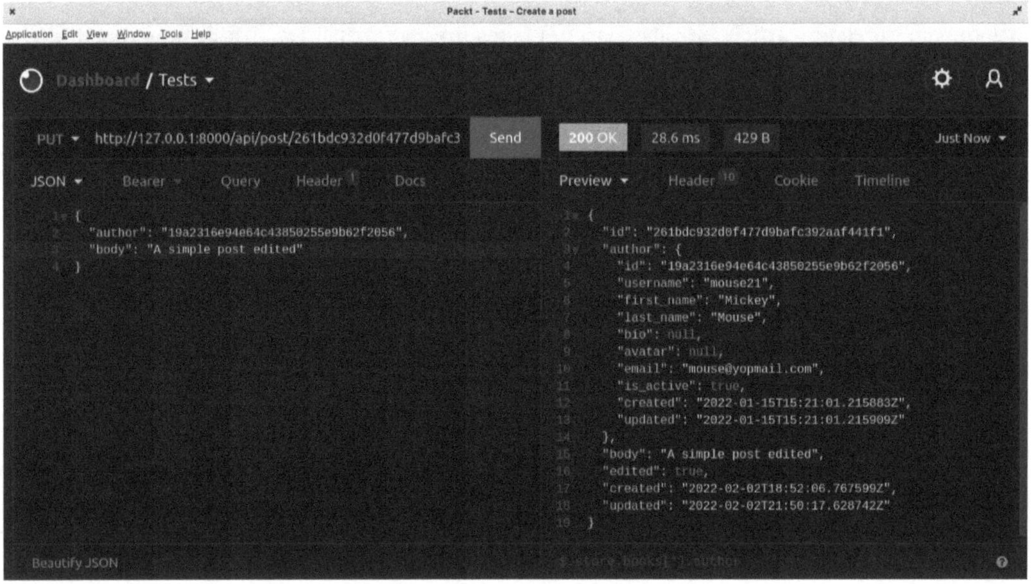

Figure 3.9 – Modifying a post

As you can see, the `edited` field in the response is set to `true`.

Let's try to delete the post and see whether it works:

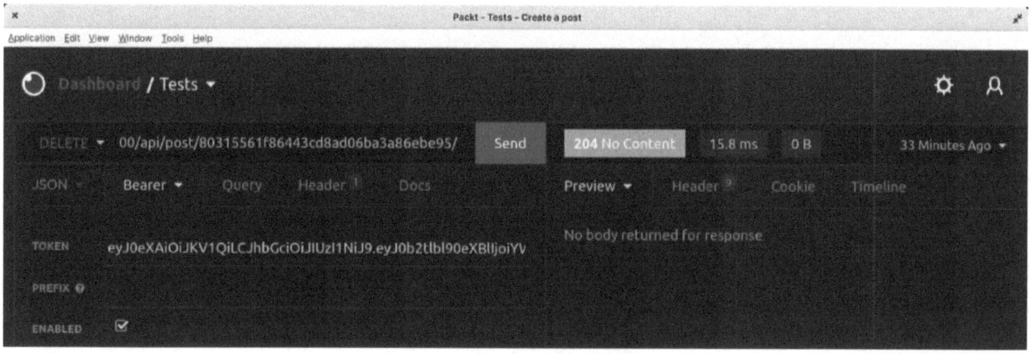

Figure 3.10 – Deleting a post

> **Important note**
>
> There is a way to delete records without necessarily deleting them from the database. It's usually called a soft delete. The record just won't be accessible to the user, but it will always be present in the database. You can learn more about this at `https://dev.to/bikramjeetsingh/soft-deletes-in-django-a9j`.

Adding the Like feature

A nice feature to have in a social media application is favoriting. Like Facebook, Instagram, or Twitter, we'll allow users here to like a post.

Plus, we'll also add data to count the number of likes a post has received and check whether a current user making the request has liked a post.

We'll do this in four steps:

1. Add a new `posts_liked` field to the `User` model.
2. Write methods on the `User` model to like and remove a like from a post. We'll also add a method to check whether the user has liked a post.
3. Add `likes_count` and `has_liked` to `PostSerializer`.
4. Add endpoints to like and dislike a post.

Great! Let's start by adding the new fields to the `User` model.

Adding the posts_liked field to the User model

The `posts_liked` field will contain all the posts liked by a user. The relationship between the `User` model and the `Post` model concerning the Like feature can be described as follows:

- A user can like many posts
- A post can be liked by many users

This kind of relationship sounds familiar? It is a *many-to-many* relationship.

Following this change, here's the updated structure of the table – we are also anticipating the methods we'll add to the model:

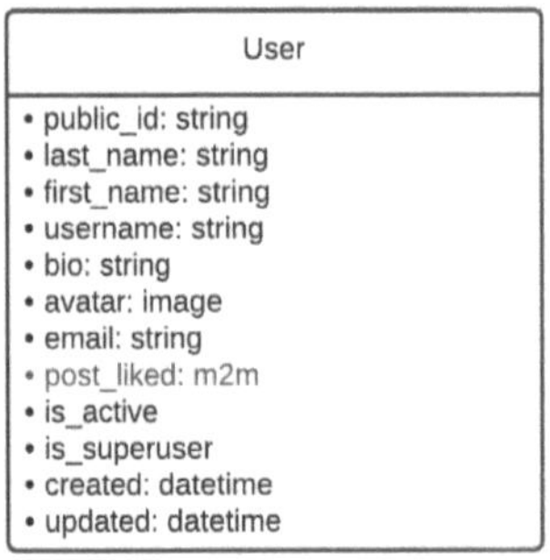

Figure 3.11 – New User table structure

Great! Let's add the `posts_liked` field to the `User` model. Open the `/core/user/models.py` file and add a new field to the `User` model:

```
class User(AbstractModel, AbstractBaseUser, PermissionsMixin):
...

    posts_liked = models.ManyToManyField(
        "core_post.Post",
        related_name="liked_by"
    )
...
```

After that, run the following commands to create a new migrations file and apply this migration to the database:

```
python manage.py makemigrations
python manage.py migrate
```

The next step is to add the new methods shown in *Figure 3.11* to the `User` model.

Adding the like, remove_like, and has_liked methods

Before writing these methods, let's describe the purpose of each new method:

- The `like()` method: This is used for liking a post if it hasn't been done yet. For this, we'll use the `add()` method from the models. We'll use `ManyToManyField` to link a post to a user.
- The `remove_like()` method: This is used for removing a like from a post. For this, we'll use the `remove` method from the models. We'll use `ManyToManyField` to unlink a post from a user.
- The `has_liked()` method: This is used for returning `True` if the user has liked a post, else `False`.

Let's move on to the coding:

```
class User(AbstractModel, AbstractBaseUser, PermissionsMixin):
    ...

    def like(self, post):
        """Like `post` if it hasn't been done yet"""
        return self.posts_liked.add(post)

    def remove_like(self, post):
        """Remove a like from a `post`"""
        return self.posts_liked.remove(post)

    def has_liked(self, post):
        """Return True if the user has liked a `post`; else
           False"""
        return self.posts_liked.filter(pk=post.pk).exists()
```

Great! Next, let's add the `likes_count` and `has_liked` fields to `PostSerializer`.

Adding the likes_count and has_liked fields to PostSerializer

Instead of adding fields such as `likes_count` in the `Post` model and generating more fields in the database, we can directly manage it on `PostSerializer`. The `Serializer` class in Django provides ways to create the `write_only` values that will be sent on the response.

Inside the `core/post/serializers.py` file, add new fields to `PostSerializer`:

Core/post/serializers.py

```
...
class PostSerializer(AbstractSerializer):
    ...
    liked = serializers.SerializerMethodField()
    likes_count = serializers.SerializerMethodField()

    def get_liked(self, instance):

        request = self.context.get('request', None)

        if request is None or request.user.is_anonymous:
            return False

        return request.user.has_liked(instance)

    def get_likes_count(self, instance):
        return instance.liked_by.count()

    class Meta:
        model = Post
        # List of all the fields that can be included in a
        # request or a response
        fields = ['id', 'author', 'body', 'edited', 'liked',
                  'likes_count', 'created', 'updated']
        read_only_fields = ["edited"]
```

In the preceding code, we are using the `serializers.SerializerMethodField()` field, which allows us to write a custom function that will return a value we want to attribute to this field. The syntax of the method will be `get_field`, where `field` is the name of the field declared on the serializer.

That is why for `liked`, we have the `get_liked` method, and for `likes_count`, we have the `get_likes_count` method.

With the new fields on `PostSerializer`, we can now add the endpoints needed to `PostViewSet` to like or dislike an article.

Adding like and dislike actions to PostViewSet

DRF provides a decorator called `action`. This decorator helps make methods on a `ViewSet` class routable. The `action` decorator takes two arguments:

- `detail`: If this argument is set to `True`, the route to this action will require a resource lookup field; in most cases, this will be the ID of the resource
- `methods`: This is a list of the methods accepted by the action

Let's write the actions on `PostViewSets`:

core/post/viewsets.py

```
...

class PostViewSet(AbstractViewSet):
    ...

    @action(methods=['post'], detail=True)
    def like(self, request, *args, **kwargs):
        post = self.get_object()
        user = self.request.user

        user.like(post)

        serializer = self.serializer_class(post)

        return Response(serializer.data,
                        status=status.HTTP_200_OK)

    @action(methods=['post'], detail=True)
    def remove_like(self, request, *args, **kwargs):
        post = self.get_object()
        user = self.request.user
```

```
        user.remove_like(post)

        serializer = self.serializer_class(post)

        return Response(serializer.data,
                        status=status.HTTP_200_OK)
```

For each action added, we are writing the logic following these steps:

1. First, we retrieve the concerned post on which we want to call the like or remove the like action. The `self.get_object()` method will automatically return the concerned post using the ID passed to the URL request, thanks to the `detail` attribute being set to `True`.
2. Second, we also retrieve the user making the request from the `self.request` object. This is done so that we can call the `remove_like` or `like` method added to the `User` model.
3. And finally, we serialize the post using the `Serializer` class defined on `self.serializer_class` and we return a response.

With this added to `PostViewSets`, the Django Rest Framework routers will automatically create new routes for this resource, and then, you can do the following:

1. Like a post with the following endpoint: `api/post/post_pk/like/`.
2. Remove the like from a post with the following endpoint: `api/post/post_pk/remove_like/`.

Great, the feature is working like a charm. In the next chapter, we'll be adding the *comments* feature to the project.

Summary

In this chapter, we've learned how to use database relationships and write permissions. We also learned how to surcharge updates and create methods on viewsets and serializers.

We performed quick refactoring on our code by creating an `Abstract` class to follow the *DRY* rule. In the next chapter, we'll be adding the Comments feature on the posts. Users will be able to create comments under posts as well as delete and update them.

Questions

1. What are some database relationships?
2. What are Django permissions?
3. How do you paginate the results of an API response?
4. How do you use Django shell?

4

Adding Comments to Social Media Posts

A social media application is more fun if your users can comment on other posts or even like them. In this chapter, we'll first learn how to add comments to posts. We'll see how we can use database relationships again to create a comment section for each post and ensure the code quality is maintained.

In this chapter, we will cover the following topics:

- Writing the Comment model
- Writing the comment serializer
- Nesting routes for the comment resource
- Writing the CommentViewSet class
- Updating a comment
- Deleting a comment

By the end of this chapter, you will be able to create Django models, write Django serializers and validation, and write nested viewsets and routes, and will have a better understanding of authorization permissions.

Technical requirements

For this chapter, you need to have Insomnia installed and some knowledge about models, database relationships, and permissions. You'll also need to have the Insomnia API client installed on your machine. The code for this chapter can be found here: `https://github.com/PacktPublishing/Full-stack-Django-and-React/tree/chap4`.

Writing the Comment model

A comment in the context of this project will represent a small text that can be viewed by anyone but only be created or updated by authenticated users. Here's what the requirements for this feature look like:

- Any user can read comments
- Authenticated users can create comments under posts
- The comment author and post author can delete comments
- The comment author can update posts

Looking at these requirements, we can definitely start with writing the model first. But first of all, let's quickly talk about the structure of the **Comment** table in the database:

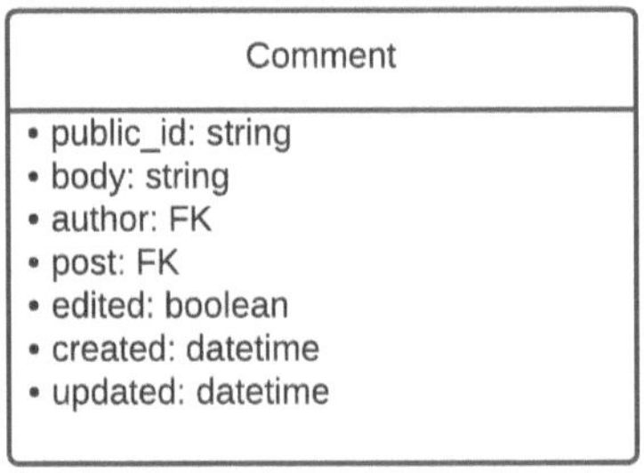

Figure 4.1 – The Comment table structure

A comment will mostly have four important fields: the author of the comment, the post on which the comment has been made, the body of the comment, and the edited field to track whether the comment has been edited or not.

As per *Figure 4.1*, we have two database relationships in the table: author and post. So, how does this schematize in the database?

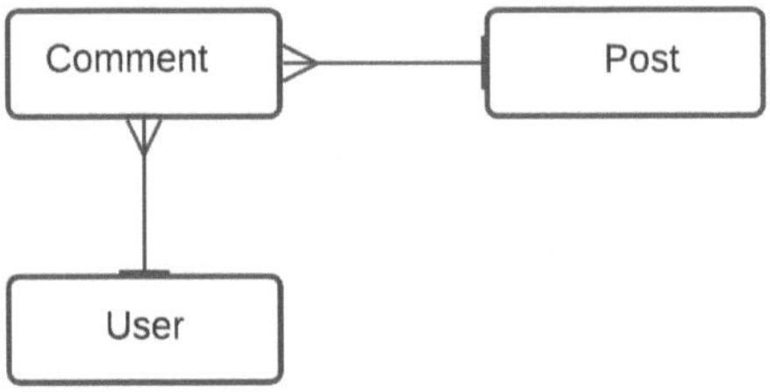

Figure 4.2 – Comment, Post, and User relationships

As you can see in *Figure 4.2*, the author (`User`) and post (`Post`) fields are **ForeignKey** types. This relates to some rules for the comment feature:

- A user can have many comments, but a comment is created by one user
- A post can have many comments, but a comment is linked to only one post

Now that we have a structure for the table and a better understanding of the requirements, let's write the model and test it.

Adding the Comment model

In `core/comment/models.py`, add the following content:

core/comment/models.py

```
from django.db import models

from core.abstract.models import AbstractModel, AbstractManager


class CommentManager(AbstractManager):
    pass


class Comment(AbstractModel):
    post = models.ForeignKey("core_post.Post",
                             on_delete=models.PROTECT)
    author = models.ForeignKey("core_user.User",
                               on_delete=models.PROTECT)

    body = models.TextField()
    edited = models.BooleanField(default=False)

    objects = CommentManager()

    def __str__(self):
        return self.author.name
```

In the preceding code snippet, we are declaring a class called `CommentManager` which is a subclass of the `AbstractManager` class. Then, we are declaring the `Comment` model class with fields such as the `post` and `author` that are respectively `ForeignKey` fields in relation to the `Post` model and the `User` model. Finally, we are declaring the body and the edited field. The rest of the code is basic formality such as telling Django with Manager class to use to manage the `Comment` model and finally a default `__str__` method to return the name of the author when checking a comment object in the Django shell.

Now that the `Comment` model is written, let's play with the model in the Django shell.

Creating a comment in the Django shell

Launch the Django shell with the following comment:

python manage.py shell

```
Python 3.10.2 (main, Jan 15 2022, 18:02:07) [GCC 9.3.0] on
linux
Type "help", "copyright", "credits" or "license" for more
information.
(InteractiveConsole)
>>> from core.comment.models import Comment
>>> from core.post.models import Post
>>> from core.user.models import User
```

First of all, we are importing the needed models to retrieve and create a comment. Next, we'll retrieve a user and a post and then write the data needed in a Python dictionary to create a comment like so:

```
>>> user = User.objects.first()
>>> post = Post.objects.first()
>>> comment_data = {"post": post, "author": user, "body": "A
comment."}
```

Now we can create the comment as follows:

```
>>> comment = Comment.objects.create(**comment_data)
>>> comment
<Comment: Dingo Dog>
>>> comment.body
'A comment.'
```

Great, now that we are sure that the comment is working, we can write the serializer for the comment feature.

Writing the comment serializer

The comment serializer will help with validation and content creation. In the comment application, create a file named `serializers.py`. We'll write `CommentSerializer` in this file.

First of all, let's import the classes and tools needed to create a serializer:

/core/comment/serializers.py

```
from rest_framework import serializers
from rest_framework.exceptions import ValidationError

from core.abstract.serializers import AbstractSerializer
from core.user.models import User
from core.user.serializers import UserSerializer
from core.comment.models import Comment
from core.post.models import Post
```

Once it's done, we can now write `CommentSerializer`:

/core/comment/serializers.py

```
...
class CommentSerializer(AbstractSerializer):
   author = serializers.SlugRelatedField(
     queryset=User.objects.all(), slug_field='public_id')
   post = serializers.SlugRelatedField(
     queryset=Post.objects.all(), slug_field='public_id')

   def to_representation(self, instance):
       rep = super().to_representation(instance)
       author =
        User.objects.get_object_by_public_id(rep["author"])
       rep["author"] = UserSerializer(author).data

       return rep

   class Meta:
       model = Comment
       # List of all the fields that can be included in a
       # request or a response
       fields = ['id', 'post', 'author', 'body', 'edited',
                 'created', 'updated']
       read_only_fields = ["edited"]
```

Let's explain the code concerning the CommentSerializer class. To create a comment, we need three fields: public_id of the author, public_id of the post, and finally, the body. We've also added validation methods for the author field.

In validate_author, we are blocking users from creating comments for other users.

And finally, the to_representation method modifies the final object by adding information about the author.

The comment serializer is now ready. We can now proceed to write the viewsets concerning the comment feature. But before that, let's talk about the endpoint of the resource.

Nesting routes for the comment resource

To create, update, or delete comments, we need to add ViewSet. In the comment directory, create a file called viewsets.py. This file will contain the code for the CommentViewSet class. We won't be writing the whole code for this viewset because we need to get some clear ideas on the structure of the endpoint.

So, add the following content for the moment:

core/comment/viewsets.py

```
from django.http.response import Http404

from rest_framework.response import Response
from rest_framework import status

from core.abstract.viewsets import AbstractViewSet
from core.comment.models import Comment
from core.comment.serializers import CommentSerializer
from core.auth.permissions import UserPermission

class CommentViewSet(AbstractViewSet):
    http_method_names = ('post', 'get', 'put', 'delete')
    permission_classes = (UserPermission,)
    serializer_class = CommentSerializer
...
```

Great, now let's talk about the endpoint architecture. The following table shows the structure of the endpoint concerning the comment. You have the method, the URL of the endpoint, and finally, the result of a call on this endpoint:

Method	URL	Result
`GET`	`/api/comment/`	Lists all the comments related to a post
`GET`	`/api/comment/comment_pk/`	Retrieves a specific comment
`POST`	`/api/comment/`	Creates a comment
`PUT`	`/api/comment/comment_pk/`	Modifies a comment
`DELETE`	`/api/comment/comment_pk/`	Deletes a comment

However, for the comment feature, we are working with posts. And it's definitely a great idea if comments are directly related to posts. Therefore, a great structure for our endpoints will look like this:

Method	URL	Action
`GET`	`/api/post/post_pk/comment/`	Lists all the comments related to a post
`GET`	`/api/post/post_pk/comment/comment_pk/`	Retrieves a specific comment
`POST`	`/api/post/post_pk/comment/`	Creates a comment
`PUT`	`/api/post/post_pk/comment/comment_pk/`	Modifies a comment
`DELETE`	`/api/post/post_pk/comment/comment_pk/`	Deletes a comment

In this structure, the endpoint is nested, meaning that comment resources live under post resources.

But how do we achieve this simply?

The Django ecosystem has a library called `drf-nested-routers`, which helps write routers to create nested resources in a Django project.

You can install this package with the following command:

```
pip install drf-nested-routers
```

Don't forget to add the dependency in the `requirements.txt` file.

Great! No need to register it in the `settings.py` file, as it doesn't come with signals, models, or applications.

In the next section, let's configure this library to fit the needs of this project.

Creating nested routes

Follow these steps to configure the `drf-nested-routers` library:

1. The first thing to do is to rewrite the `routers.py` file:

core/routers.py

```
from rest_framework_nested import routers
...
router = routers.SimpleRouter()
...
```

`drf-nested-routers` comes with an extended `SimpleRouter`, which will be useful for creating nested routes.

2. After that, create a new nested route called `POST`:

```
...
# ##########################################################
############### #
# ####################
POST                        ######################## #
# ##########################################################
############### #
router.register(r'post', PostViewSet, basename='post')

posts_router = routers.NestedSimpleRouter(router,
r'post', lookup='post')
...
```

`NestedSimpleRouter` is a sub-class of the `SimpleRouter` class, which takes initialization parameters, such as `parent_router` – `router` -`parent_prefix` – `r'post'` – and the lookup – `post`. The lookup is the regex variable that matches an instance of the parent resource – `PostViewSet`.

In our case, the lookup regex will be `post_pk`.

3. The next step is to register the `comment` route on `post_router`:

core/routers.py

```
...
# ##############################################################
############### #
# ####################
POST                           ######################## #
# ##############################################################
############### #
router.register(r'post', PostViewSet, basename='post')

posts_router = routers.NestedSimpleRouter(router,
r'post', lookup='post')
posts_router.register(r'comment', CommentViewSet,
basename='post-comment')
urlpatterns = [
   *router.urls,
   *posts_router.urls
]
...
```

Great! The comment resource is available, but we must rewrite the `create`, `get_object`, and `get_queryset` methods on the `CommentViewSet` class. Let's see how using nested routes can modify the logic of retrieving objects in the next section.

Writing the CommentViewSet class

We now have a clear idea of how the endpoint will work.

Follow these steps in the `core/comment/viewsets.py` file to finish writing the `CommentViewSet` class:

1. Rewrite the `get_queryset` method of the `CommentViewSet` class to fit the new architecture of the endpoint:

core/comment/viewsets.py

```
...
class CommentViewSet(AbstractViewSet):
...
   def get_queryset(self):
```

```
        if self.request.user.is_superuser:
            return Comment.objects.all()


        post_pk = self.kwargs['post_pk']
        if post_pk is None:
            return Http404
        queryset = Comment.objects.filter(
          post__public_id=post_pk)


        return queryset
```

In the preceding code, `get_queryset` is the method called when the user hits the `/api/post/post_pk/comment/` endpoint. The first verification here is to check whether the user is a superuser. If that's the case, we return all the comment objects in the database.

If the user is not a superuser, then we'll return the comments concerning a post. With the post nested route, we set the `lookup` attribute to `post`. That means that in `kwargs` (a dictionary containing additional data) of every request, a public id value of the `post` with the dictionary key `post_pk` will be passed in the URL of the endpoint.

If that's not the case, we just return a 404 Not Found response.

We then make a query to the database by filtering and retrieving only comments that have the `post.public_id` field equal to `post_pk`. This is done with the filter method provided by the Django ORM. It's useful to write conditions for retrieving objects from the database.

2. Next, let's add the `get_object` method to the same `CommentViewSet` so we can use the `public_id` to retrieve the specific comment:

core/comment/viewsets.py

```
...
class CommentViewSet(AbstractViewSet):
...
    def get_object(self):
        obj = Comment.objects.get_object_by_public_id(
          self.kwargs['pk'])
        self.check_object_permissions(self.request,
                                      obj)
        return obj
...
```

Similar to the UserViewSet get_object method, this method is called on each request made to the /api/post/post_pk/comment/comment_pk/ endpoint. Here, pk is represented by comment_pk.

Then, we retrieve the object and check for permissions. If everything is good, we return the object.

3. And as the last step, let's write the create method:

core/comment/viewsets.py

```
...
class CommentViewSet(AbstractViewSet):
...
    def create(self, request, *args, **kwargs):
        serializer =
          self.get_serializer(data=request.data)
        serializer.is_valid(raise_exception=True)
        self.perform_create(serializer)
        return Response(serializer.data,
                        status=status.HTTP_201_CREATED)
```

Similar to the create method on PostViewSet, we pass request.data to the ViewSet serializer – CommentSerialier – and try to validate the serializer.

If everything is good, we move to create a new object – a new comment – based on the serializer from CommentSerializer.

Great! We now have a fully functional ViewSet. Next, let's test the features with Insomnia.

Testing the comments feature with Insomnia

Before trying to retrieve comments, let's create some comments with POST on the /api/post/post_id/comment/ URL by following these steps:

1. Replace post_id with public_id of a post that you have already created.

   Here's an example of a payload for this request:

```
{
    "author": "61c5a1ecb9f5439b810224d2af148a23",
    "body": "Hey! I like your post.",
    "post": "e2401ac4b29243e6913bd2d4e0944862"
}
```

And here's a screenshot of a request made to create a comment in Insomnia:

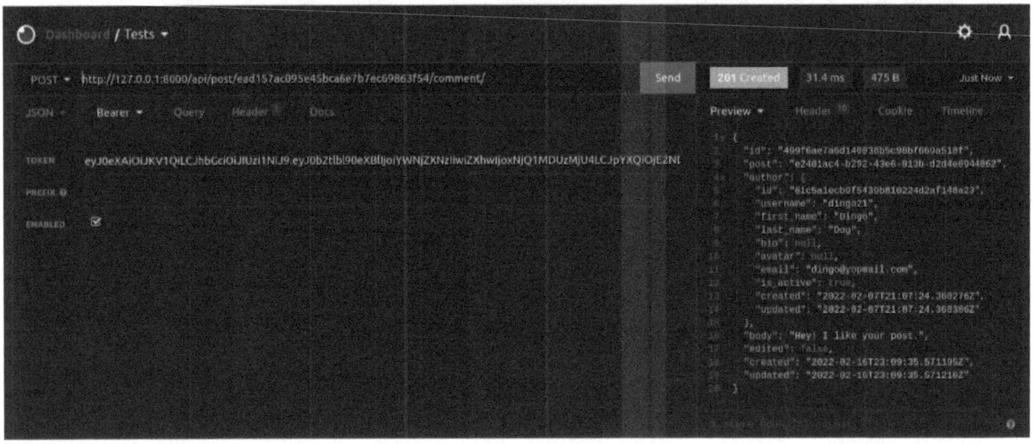

Figure 4.3 – Creating a post

2. Great! Now, modify the type of request from `POST` to `GET`. You'll get all the comments concerning the post:

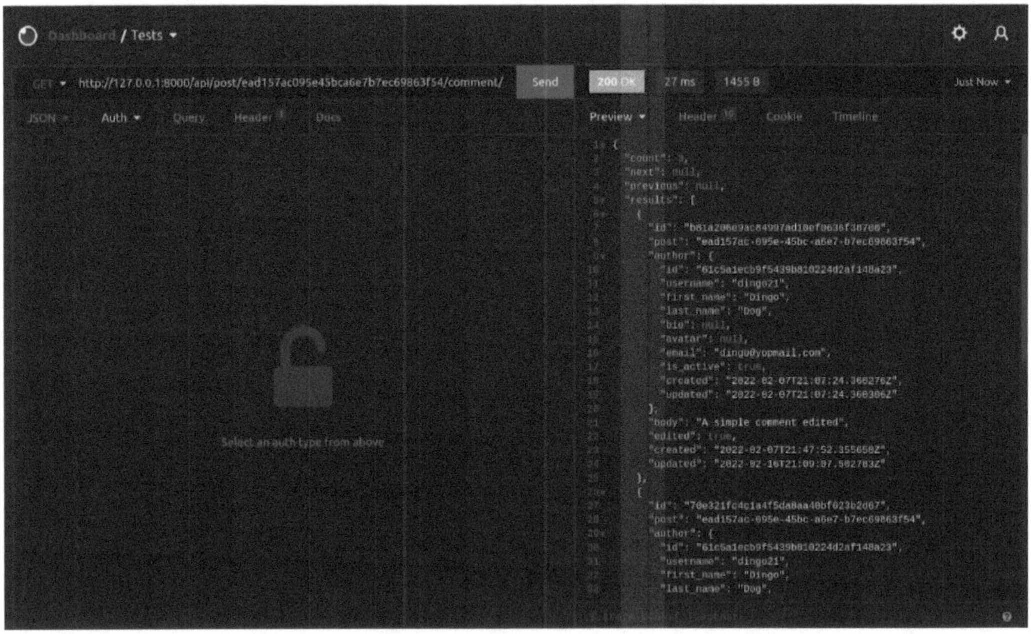

Figure 4.4 – Listing all comments

Now that it's possible to create a comment without issues, let's add a feature for updating a comment and deleting a comment.

Updating a comment

Updating a comment is an action that can only be done by the author of the comment. And the user should only be able to update the body field of the comment and can't modify the author value. Follow these steps to add the update feature:

1. In `core/comment/viewsets`, make sure that `put` is in the list of `http_method_names` of `CommentViewSet`:

core/comment/viewsets

```
...
class CommentViewSet(AbstractViewSet):
    http_method_names = ('post', 'get', 'put',
                         'delete')
...
```

After that, let's write a `validate` method for the `post` field. We want to make sure that this value is not editable on `PUT` requests.

2. Inside the `core/comment/serializers.py` file, add a new method called `validate_post` to `CommentSerializer`:

core/comment/serializers.py

```
...
def validate_post(self, value):
    if self.instance:
        return self.instance.post
    return value
...
```

Every model serializer provides an `instance` attribute that holds the object that will be modified if there is a `delete`, `put`, or `patch` request. If this is a `GET` or `POST` request, this attribute is set to `None`.

3. Next, let's rewrite the `update` method on the `CommentSerializer` class. We'll rewrite this class to pass the edited value to `True`:

core/comment/serializers.py

```
...
class CommentSerializer(AbstractSerializer):
    ...
    def update(self, instance, validated_data):
        if not instance.edited:
            validated_data['edited'] = True
        instance = super().update(instance,
                                  validated_data)
        return instance
...
```

4. Great! Now, let's try a PUT request in Insomnia on the `/api/post/post_pk/comment/comment_pk/` endpoint. Here's an example of a JSON body for the request:

```
{
    "author": "61c5a1ecb9f5439b810224d2af148a23",
    "body": "A simple comment edited",
    "post": "e2401ac4b29243e6913bd2d4e0944862"
}
```

And here's a screenshot of a PUT request in Insomnia:

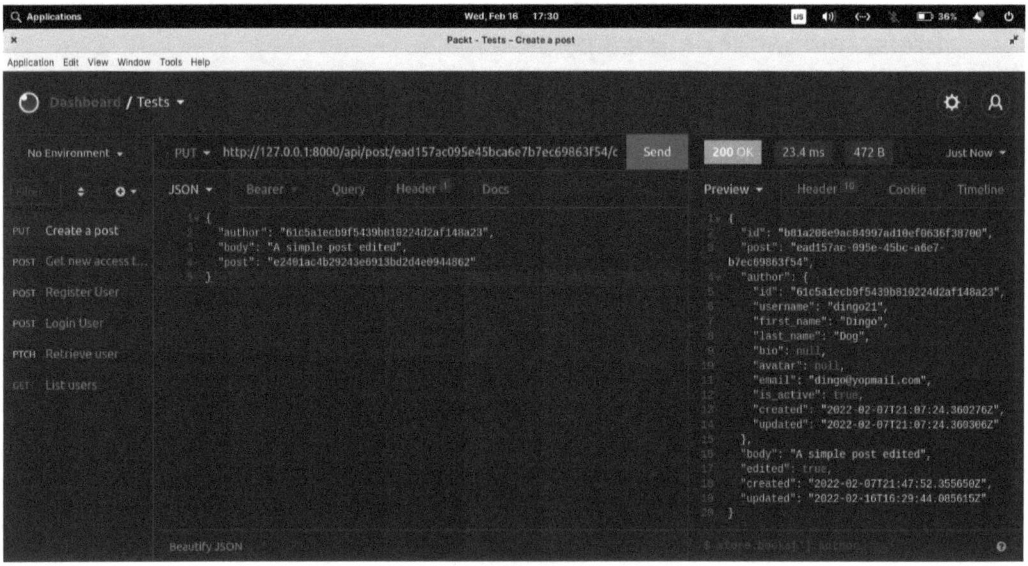

Figure 4.5 – Modifying a post

You will notice in the response body that the `edited` field is set to `true`, and the body of the comment has changed as well.

Now that it's possible to modify a comment, let's add the feature for deleting a comment.

Deleting a comment

Deleting a comment is an action that can only be performed by the author of the post, the author of the comment, and a superuser. To implement this rule, we'll simply add some permissions in the `UserPermission` class by following these steps:

1. Make sure that `delete` is in the list of `http_method_names` of the `CommentViewSet` class:

core/comment/viewsets

```
...
class CommentViewSet(AbstractViewSet):
    http_method_names = ('post', 'get', 'put',
                         'delete')
...
```

2. Once it's done, let's add more verifications in the `core/auth/permissions` file in the `has_object_permission` method of the `UserPermission` class:

core/auth/permissions

```
...
def has_object_permission(self, request, view, obj):
...


    if view.basename in ["post-comment"]:
        if request.method in ['DELETE']:
            return bool(request.user.is_superuser or
                        request.user in [obj.author,
                        obj.post.author])


        return bool(request.user and
                    request.user.is_authenticated)
...
```

All requests can be made on the `post-comment` endpoint. However, if the method of the request is `DELETE`, we check whether the user is a superuser, the author of the comment, or the author of the post.

3. Let's try to delete the comment in Insomnia at this endpoint: `/api/post/post_pk/comment/comment_pk/`. Make sure you have the access token of the post author or the comment author.

   And here's a screenshot of a `DELETE` request to delete a comment under a post:

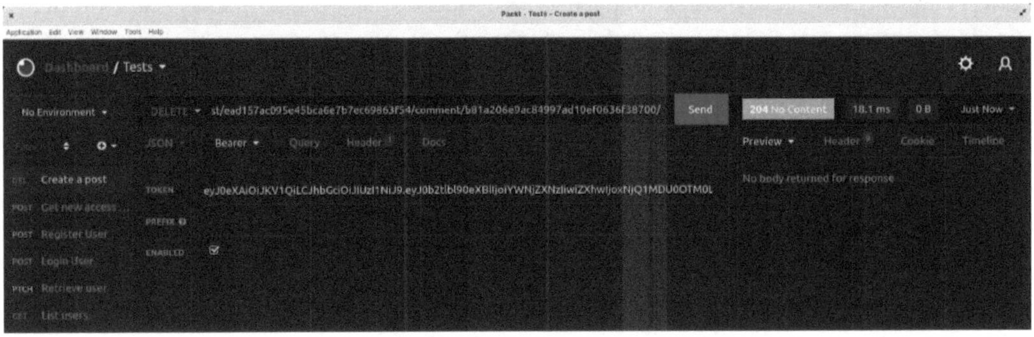

Figure 4.6 – Deleting a post

Great, the feature is working like a charm. And we've just learned how to write permissions for a `DELETE` request.

Summary

In this chapter, we've learned how to create a comment feature for the posts in our social media projects. That led us to learn more about how to better structure an endpoint using nested routers but also how to write custom permissions.

We've also dived deeper into serializer validations and how they work on different HTTP requests.

In the next chapter, we'll focus on writing unit and integration tests for every feature added to the project.

Questions

1. What is a nested route?
2. What is `drf-nested-routers`?
3. Which attribute on a model serializer can help you to know whether the request is a `PUT` or a `DELETE` request?

5
Testing the REST API

In software engineering, testing is a process to check whether the actual software product performs as expected and is bug free.

There are a lot of ways to test software through both manual and automated tests. But in this project, we'll focus more on **automated testing**. However, we'll first dive into the different ways of testing software, including their pros and their cons, and also talk about the concept of the **testing pyramid**. We'll also check the tools needed to add tests to a Django application and add tests to the models and the viewsets. This chapter will help you understand testing for developers and also how to write tests for a Django API.

In this chapter, we'll be covering the following topics:

- What is testing?
- Testing in Django
- Configuring the testing environment
- Writing tests for Django models
- Writing tests for Django viewsets

Technical requirements

You can find the code of the current chapter at this link: `https://github.com/PacktPublishing/Full-stack-Django-and-React/tree/chap5`.

What is testing?

To make it simple, testing is finding out how well something works.

However, the process comprises a group of techniques to determine the correctness of the application under a script or manual test directly on the user interface. The aim is to detect failures, including bugs and performance issues, in the application, so that they can be corrected.

Most of the time, testing is done by comparing the software requirements to the actual software product. If one of the requirements is to make sure that input only accepts numbers and not characters or files, a test will be conducted to check whether the input has a validation system to reject non-number values in the input.

However, testing also involves an examination of code and the execution of code in various environments and conditions.

What is software testing?

Software testing is the process of examining the behavior of the software under test for validation or verification. It considers the attributes of reliability, scalability, reusability, and usability to evaluate the execution of the software components (servers, database, application, and so on) and find software bugs, errors, or defects.

Software testing has a lot of benefits, some of which are as follows:

- **Cost effectiveness**: Testing any software project helps the business save money in the long run. As the process helps detect bugs and check whether newly added features are working in the system without breaking things, it's a great technical debt reducer.
- **Security**: If testing is done well, it can be a quick way to detect security risks and problems at an early stage before deploying a product to the whole world.
- **Product quality**: Testing helps with performance measurement, making sure that the requirements are respected.

Why is software testing important?

Testing your software is important because it helps reduce the impact of bugs through bug identification and resolution. Some bugs can be quite dangerous and can lead to financial losses or endanger human life. Here are some historical examples:

Source: `https://lexingtontechnologies.ng/software-testing/`.

- In April 1999, $1.2 billion were lost due to the failure of a military satellite launch. To date, this is the costliest accident in the history of the world.
- In 2014, the giant Nissan recalled over 1 million cars from the market because of a software failure in the airbag sensory detectors.
- In 2014, some of Amazon's third-party retailers lost a lot of money because of a software glitch. The bug affected the price of the products, reducing them to 1p.
- In 2015, a software failure in the **Point of sales** (**POS**) system of Starbucks stores caused the temporary closure of more than 60% of their stores in the US and Canada.

- In 2015, an F-35 fighter plane fell victim to a software bug, which prevented it from detecting or identifying targets correctly. The sensor on the plane was unable to identify threats even from their own planes.
- In 2016, Google reported a bug affecting Windows 10 machines. The vulnerability allowed users to escape security sandboxes through a flow in the win32k system.

What are the various types of testing?

Testing is typically classified into three categories:

- **Functional testing**: This type of testing comprises unit, integration, user acceptance, globalization, internationalization testing, and so on
- **Non-functional testing**: This type of testing checks for factors such as performance, volume, scalability, usability, and load
- **Maintenance testing**: This type of testing considers regression and maintenance

However, these tests can also be classified into two different types:

- Automated tests
- Manual tests

First, let's see what manual testing is.

Understanding manual testing

Manual testing is the process of testing software manually to find defects or bugs. It's the process of testing the functionalities of an application without the help of automation tools.

An example of manual testing is when test users are called to test an application or a special feature. They can be asked to test a specific form, push the application to its limits when it comes to performance, and much more.

Manual testing has a lot of advantages:

- It's very useful to test user interface designs and interactions
- It's easier to learn for new testers
- It takes user experience and usability into consideration
- It's cost-effective

However, manual testing also has some cons:

- It requires human resources.
- It's time-consuming.
- Testers consider test cases based on their skills and experience. This means that a beginner tester may not cover all the functions.

Even if manual testing sounds very appealing, it can be quite a time- and resource-consuming exercise, and developers definitely do not make really good manual testers. Let's see how automated testing can erase the cons of manual testing and place better development at the center of testing.

Understanding automated testing

Automated testing is simply the process of testing software using automation tools to find defects. These automation tools can be scripts written in the language used to build the application or some software or drivers (such as **Selenium**, **WinRunner**, and **LoadRunner**) to make automated testing easier and faster.

Automated testing fixes the cons of manual testing, and it also has more advantages, as shown in the following list:

- Faster in execution
- Cheaper than manual testing in the long run
- More reliable, powerful, and versatile
- Very useful in regression testing
- Able to provide better test coverage
- Possible to run without human intervention
- Much cheaper

However, automated testing is also inconvenient in some ways:

- It is expensive at the beginning
- It has a huge cost of maintenance when requirements change
- Automated testing tools are expensive

The real value of automated testing and manual testing comes when each is used in the right environment.

For example, manual testing is much more useful on frontend projects where you want to test the usability and user experience. Automated testing can be useful to test methods or functions in the code and is very useful for finding bugs or security issues.

In this chapter, we'll focus on writing automated tests in Python. As we are developing an API, we want to make sure that the system is reliable and behaves as we want it to, but it should also be secure against the possible issues of the next added feature.

This said, let's talk about testing in Django and introduce the notion of **test-driven development** (TDD).

Testing in Django

Testing in Python, particularly in Django, is very simple and easy. The framework actually provides many tools and utilities you can use to write tests for the models, serializers, or views in the application.

However, the Python ecosystem for testing relies a lot on one tool to write tests, and this tool has deep integration with Django. The tool is named **Pytest** (`https://docs.pytest.org`) and is a framework for writing small and readable tests. Used with Django, Pytest is mainly used for API testing by writing code to test API endpoints, databases, and user interfaces.

But why use Pytest? Well, it has the following advantages:

- It is free and open source
- It has a simple syntax and is very easy to start with
- It automatically detects test files, functions, and classes
- It can run multiple tests in parallel, increasing the performance and the speed of running tests

We'll use Pytest in this project to write two kinds of tests: **integration tests** and **unit tests**.

Before starting to code, let's learn about integration testing and unit testing by considering the concepts of TDD and the testing pyramid.

The testing pyramid

The testing pyramid is a framework that can help developers start with testing to create high-quality software. Basically, the testing pyramid specifies the types of tests that should be included in an automated test suite.

First of all, remember that the testing pyramid operates at three levels:

- Unit tests
- Integration tests
- End-to-end tests

The following figure shows the positions of each of these levels in the pyramid and how they are prioritized in terms of the speed performance and level of integration or isolation:

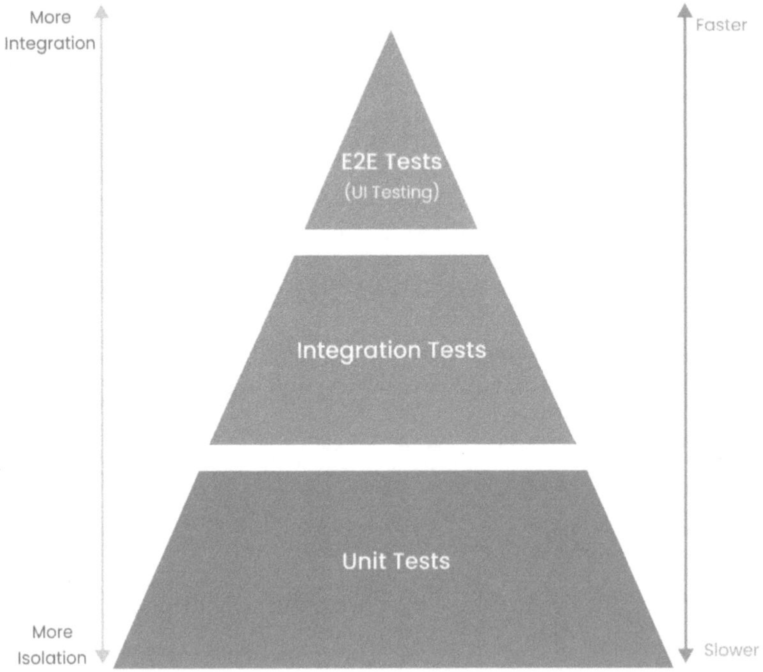

Figure 5.1 – The testing pyramid

In the preceding figure, the base level is occupied by unit testing. Unit tests target individual components or functionality to check whether they work as expected in isolated conditions. In our backend project, an example would be to test whether the `like_post` method on the `User` class model actually performs as intended. We are not testing the whole `User` model; we are testing one method of the `User` model class.

It's definitely a good habit to write a lot of unit tests. They should comprise at least 60% of all the tests in your code base because they are fast, short, and test a lot of components.

On the second level, you have integration tests. If unit tests verify small pieces of a code base, integration tests test how this code interacts with other code or other parts of the software. A useful, albeit controversial, example of integration testing is writing a test for a viewset. When testing a viewset, you are also testing the permissions, the authentication classes, the serializers, the models, and the database if possible. It's a test of how the different parts of the Django API work together.

An integration test can also be a test between your application and an external service, a payment API, for example.

On the third level at the top of the pyramid, you have end-to-end tests. These kinds of tests ensure that the software is working as required. They test how the application works from beginning to end.

In this book, we'll focus on unit and integration testing. Note that integration tests are the subject of some misunderstandings that will be cleared up once we define them. According to my personal experience, unit tests in Django are written more on the model and serializer side of each application. They can be used for testing the creation of an object in the database as well as for retrieving, updating, or deletion.

Regarding viewset tests, I believe that they can act as integration tests because running them calls on permissions, authentication, serializers, validation, and also models, depending on the action you are performing.

Returning to unit tests, they are more effective when using TDD, which comprises software development practices that focus on writing unit test cases before developing the feature. Even if it sounds counter-intuitive, TDD has a lot of advantages:

- It ensures optimized code
- It ensures the application of design patterns and better architecture
- It helps the developer understand the business requirements
- It makes the code flexible and easier to maintain

However, we didn't particularly respect the TDD rule in the book. We relied on the Django shell and a client to test the feature of the REST API we are building. For the next features that will be added to the project, tests will be written before coding the feature.

With concepts such as TDD, unit and integration testing, and testing pyramid understood, we can now configure the testing environment.

Configuring the testing environment

Pytest, taken alone, is simply a Python framework to write unit tests in Python programs. Thankfully, there is a plugin for Pytest to write tests in Django projects and applications.

Let's install and configure the environment for testing by using the following command:

```
pip install pytest-django
```

Once the package is installed, create a new file called `pytest.ini` at the root of the Django project:

pytest.ini

```
[pytest]
DJANGO_SETTINGS_MODULE=CoreRoot.settings
python_files=tests.py test_*.py *_tests.py
```

Once it's done, run the `pytest` command:

```
pytest
```

You'll see the following output:

```
========================== test session starts
==============================
platform linux -- Python 3.10.2, pytest-7.0.1, pluggy-1.0.0
django: settings: CoreRoot.settings (from ini)
rootdir: /home/koladev/PycharmProjects/Full-stack-Django-and-
React, configfile: pytest.ini
plugins: django-4.5.2
collected 0 items
```

Great! Pytest is installed in the project, and we can write the first test in the project to test the configuration.

Writing your first test

The Pytest environment is configured, so let's see how we can write a simple test using Pytest.

At the root of the project, create a file called `tests.py`. We'll simply write a test to test the sum of a function.

Following the TDD concept, we'll write the test first and make it fail:

tests.py

```
def test_sum():
    assert add(1, 2) == 3
```

This function is written to check for a condition, justifying the usage of the `assert` Python keyword. If the condition after the `assert` is true, the script will continue or stop the execution. If that's not the case, an assertion error will be raised.

If you run the `pytest` command, you'll receive the following output:

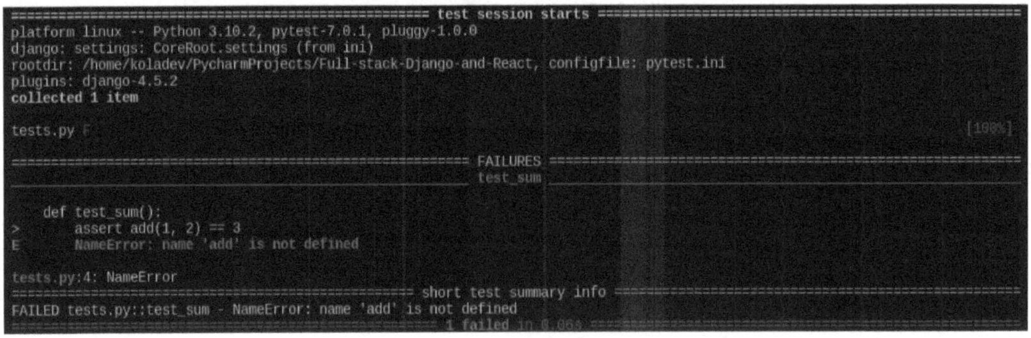

Figure 5.2 – Failing tests

From the preceding output, we are sure that the test has failed. Let's now write the feature to pass the test.

In the same file, `tests.py`, add the following function:

tests.py

```
def add(a, b):
    return a + b


def test_sum():
    assert sum(1, 2) == 3
```

Now, run the `pytest` command again in the terminal. Everything should now be green:

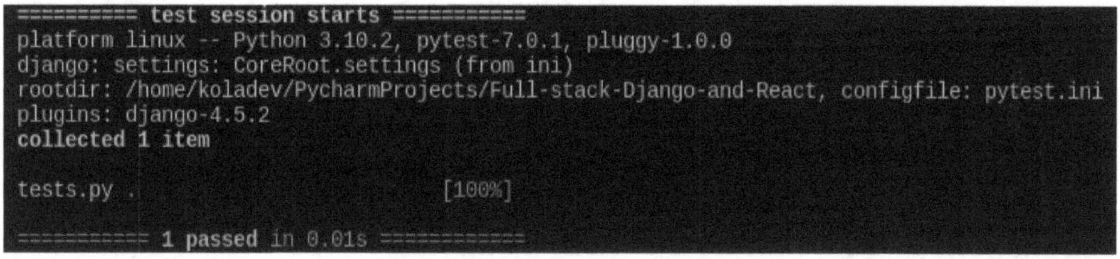

Figure 5.3 – Test passes successfully

Great! You have written the first test in the project using Pytest. In the next section, we'll be writing tests for the models of the project.

Writing tests for Django models

When applying testing to a Django project, it's always a good idea to start with writing tests for the models. But why test the models?

Well, it gives you better confidence in your code and the connections to the database. It'll make sure that methods or attributes on the model are well represented in the database, but it can also help you with better code structure, resolving bugs, and building documentation.

Without further ado, let's start by writing tests for the `User` model.

Writing tests for the User model

Inside the `core/user` directory, create a new file called `tests.py`. We'll write tests to create a user and a simple user:

core/user/tests.py

```
import pytest
from core.user.models import User

data_user = {
    "username": "test_user",
    "email": "test@gmail.com",
    "first_name": "Test",
    "last_name": "User",
    "password": "test_password"
}
```

Once the imports and the data to create the user have been added, we can write the test function:

core/user/tests.py

```
@pytest.mark.django_db
def test_create_user():
    user = User.objects.create_user(**data_user)
    assert user.username == data_user["username"]
    assert user.email == data_user["email"]
    assert user.first_name == data_user["first_name"]
    assert user.last_name == data_user["last_name"]
```

Above the `test_create_user` function, you'll probably notice some syntax. It's called a decorator, and it's basically a function that takes another function as its argument and returns another function.

`@pytest.mark.django_db` gives us access to the Django database. Try to remove this decorator and run the tests.

You'll get an error output with a similar message at the end:

```
==============================================================
==== short test summary info =================================
================================
FAILED core/user/tests.py::test_create_user - RuntimeError:
Database access not allowed, use the "django_db" mark, or the
"db" or "transactional_db" fixture...
```

Well, re-add the decorator and run the pytest command and all tests should pass normally.

Let's do another test to make sure that the creation of superuser works perfectly.

Add a new dictionary containing the data needed to create superuser:

core/user/tests.py

```
data_superuser = {
    "username": "test_superuser",
    "email": "testsuperuser@gmail.com",
    "first_name": "Test",
    "last_name": "Superuser",
    "password": "test_password"
}
```

And here's the function that tests the creation of superuser:

core/user/tests.py

```
@pytest.mark.django_db
def test_create_superuser():
    user = User.objects.create_superuser(**data_superuser)
    assert user.username == data_superuser["username"]
    assert user.email == data_superuser["email"]
    assert user.first_name == data_superuser["first_name"]
    assert user.last_name == data_superuser["last_name"]
    assert user.is_superuser == True
    assert user.is_staff == True
```

Run the tests again, and everything should be green.

Great! Now that we have a better understanding of how pytest works for tests, let's write tests for the Post model.

Writing tests for the Post model

To create a model, we need to have a user object ready. This will also be the same for the Comment model. To avoid repetition, we'll simply write **fixtures**.

A fixture is a function that will run before each test function to which it's applied. In this case, the fixture will be used to feed some data to the tests.

To add fixtures in the project, create a new Python package called `fixtures` in the `core` directory.

In the `core/fixtures` directory, create a file called `user.py`. This file will contain a user fixture:

core/fixtures/user.py

```
import pytest
from core.user.models import User

data_user = {
    "username": "test_user",
    "email": "test@gmail.com",
    "first_name": "Test",
    "last_name": "User",
    "password": "test_password"
}

@pytest.fixture
def user(db) -> User:
    return User.objects.create_user(**data_user)
```

In the preceding code, the `@pytest.fixture` decorator labels the function as a fixture. We can now import the `user` function in any test and pass it as an argument to the `test` function.

Inside the `core/post` directory, create a new file called `tests.py`. This file will then test for the creation of a post.

Here's the code:

core/post/tests.py

```
import pytest

from core.fixtures.user import user
from core.post.models import Post

@pytest.mark.django_db
def test_create_post(user):
    post = Post.objects.create(author=user,
                               body="Test Post Body")
    assert post.body == "Test Post Body"
    assert post.author == user
```

As you can see, we are importing the `user` function from `user.py` in the fixtures directory and passing it as an argument to the `test_create_post` test function.

Run the `pytest` command, and everything should be green.

Now that we have a working test for the `Post` model, let's write tests for the `Comment` model.

Writing tests for the Comment model

Writing tests for the `Comment` model requires the same steps as the tests for the `Post` model. First of all, create a new file called `post.py` in the `core/fixtures` directory.

This file will contain the fixture of a post, as it's needed to create a comment.

But the `post` fixture will also need a `user` fixture. Thankfully, it's possible with Pytest to inject fixtures into other fixtures.

Here's the code for the `post` fixture:

core/fixtures/post.py

```
import pytest

from core.fixtures.user import user
from core.post.models import Post

@pytest.fixture
def post(db, user):
    return Post.objects.create(author=user,
                               body="Test Post Body")
```

Great! With the fixtures added, we can now write the test for comment creation.

Inside the `core/comment/` directory, create a new file called `tests.py`:

core/comment/tests.py

```
import pytest

from core.fixtures.user import user
from core.fixtures.post import post
from core.comment.models import Comment
```

```
@pytest.mark.django_db
def test_create_comment(user, post):
    comment = Comment.objects.create(author=user, post=post,
      body="Test Comment Body")
    assert comment.author == user
    assert comment.post == post
    assert comment.body == "Test Comment Body"
```

Run the tests with the `pytest` command, and everything should be green.

Great! We've just written tests for all the models in the project. Let's move on to writing tests for the viewsets.

Writing tests for your Django viewsets

Viewsets or endpoints are the interfaces of the business logic that the external clients will use to fetch data and create, modify, or delete data. It's always a great habit to have tests to make sure that the whole system, starting from a request to the database, is working as intended.

Before starting to write the tests, let's configure the Pytest environment to use the API client from DRF.

The API client is a class that handles different HTTP methods, as well as features such as authentication in testing, which can be very helpful for directly authenticating without a username and password to test some endpoints. Pytest provides a way to add configurations in a testing environment.

Create a file named `conftest.py` at the root of the project. Inside the file, we'll create a fixture function for our custom client:

conftest.py

```
import pytest
from rest_framework.test import APIClient

@pytest.fixture
def client():
    return APIClient()
```

Great! We can now directly call this client in the next tests.

Let's start by testing the authentication endpoints.

Writing tests for authentication

Inside the `core/auth` directory, create a file named `tests.py`. Instead of writing test functions directly, we write a class that will contain the testing methods as follows:

core/auth/tests.py

```
import pytest
from rest_framework import status

from core.fixtures.user import user

class TestAuthenticationViewSet:

    endpoint = '/api/auth/'
```

Let's add the `test_login` method to the `TestAuthenticationViewSet` class:

Core/auth/tests.py

```
...
    def test_login(self, client, user):
        data = {
            "username": user.username,
            "password": "test_password"
        }
        response = client.post(self.endpoint + "login/",
                               data)

        assert response.status_code == status.HTTP_200_OK
        assert response.data['access']
        assert response.data['user']['id'] ==
          user.public_id.hex
        assert response.data['user']['username'] ==
          user.username
        assert response.data['user']['email'] == user.email
   ...
```

This method basically tests the login endpoint. We are using the client fixture initialized in the `conftest.py` file to make a `post` request. Then, we test for the value of `status_code` of the response and the response returned.

Run the `pytest` command, and everything should be green.

Let's add tests for the `register` and `refresh` endpoints:

core/auth/tests.py

```
...
    @pytest.mark.django_db
    def test_register(self, client):
        data = {
            "username": "johndoe",
            "email": "johndoe@yopmail.com",
            "password": "test_password",
            "first_name": "John",
            "last_name": "Doe"
        }

        response = client.post(self.endpoint + "register/",
                               data)
        assert response.status_code ==
          status.HTTP_201_CREATED

    def test_refresh(self, client, user):

       data = {
            "username": user.username,
            "password": "test_password"
        }
        response = client.post(self.endpoint + "login/",
                                 data)
        assert response.status_code == status.HTTP_200_OK

        data_refresh = {
            "refresh":  response.data['refresh']
        }

        response = client.post(self.endpoint + "refresh/",
                               data_refresh)
        assert response.status_code == status.HTTP_200_OK
        assert response.data['access']
```

In the preceding code, within the `test_refresh` method, we log in to get a refresh token to make a request to get a new access token.

Run the `pytest` command again to run the tests, and everything should be green.

Let's move on to writing tests for `PostViewSet`.

Writing tests for PostViewSet

Before starting to write the viewsets tests, let's quickly refactor the code to simply write the tests and follow the DRY rule. Inside the `core/post` directory, create a Python package called `tests`. Once it's done, rename the `tests.py` file in the `core/post` directory to `test_models.py` and move it to the `core/post/tests/` directory.

Inside the same directory, create a new file called `test_viewsets.py`. This file will contain tests for `PostViewSet`:

core/post/tests/test_viewsets.py

```
from rest_framework import status

from core.fixtures.user import user
from core.fixtures.post import post

class TestPostViewSet:

    endpoint = '/api/post/'
```

`PostViewSet` handles requests for two types of users:

- Authenticated users
- Anonymous users

Each type of user has different permissions on the `post` resource. So, let's make sure that these cases are handled:

core/post/tests/test_viewsets.py

```
...
    def test_list(self, client, user, post):
        client.force_authenticate(user=user)
        response = client.get(self.endpoint)
        assert response.status_code == status.HTTP_200_OK
        assert response.data["count"] == 1
```

```
    def test_retrieve(self, client, user, post):
        client.force_authenticate(user=user)
        response = client.get(self.endpoint +
                              str(post.public_id) + "/")
        assert response.status_code == status.HTTP_200_OK
        assert response.data['id'] == post.public_id.hex
        assert response.data['body'] == post.body
        assert response.data['author']['id'] ==
          post.author.public_id.hex
```

For these tests, we are forcing authentication. We want to make sure that authenticated users have access to the post's resources. Let's now write a test method for post creation, updating, and deletion:

core/post/tests/test_viewsets.py

```
...
    def test_create(self, client, user):
        client.force_authenticate(user=user)
        data = {
            "body": "Test Post Body",
            "author": user.public_id.hex
        }
        response = client.post(self.endpoint, data)
        assert response.status_code ==
          status.HTTP_201_CREATED
        assert response.data['body'] == data['body']
        assert response.data['author']['id'] ==
          user.public_id.hex

    def test_update(self, client, user, post):
        client.force_authenticate(user=user)
        data = {
            "body": "Test Post Body",
            "author": user.public_id.hex
        }
        response = client.put(self.endpoint +
          str(post.public_id) + "/", data)

        assert response.status_code == status.HTTP_200_OK
```

```
        assert response.data['body'] == data['body']

    def test_delete(self, client, user, post):
        client.force_authenticate(user=user)
        response = client.delete(self.endpoint +
          str(post.public_id) + "/")
        assert response.status_code ==
          status.HTTP_204_NO_CONTENT
```

Run the tests, and the outcomes should be green. Now, for the anonymous users, we want them to access the resource in reading mode, so they can't create, modify, or delete a resource. Let's test and validate these features:

core/post/tests/test_viewsets.py

```
...
    def test_list_anonymous(self, client, post):
        response = client.get(self.endpoint)
        assert response.status_code == status.HTTP_200_OK
        assert response.data["count"] == 1

    def test_retrieve_anonymous(self, client, post):
        response = client.get(self.endpoint +
          str(post.public_id) + "/")
        assert response.status_code == status.HTTP_200_OK
        assert response.data['id'] == post.public_id.hex
        assert response.data['body'] == post.body
        assert response.data['author']['id'] ==
          post.author.public_id.hex
```

Run the tests to make sure everything is green. After that, let's test the forbidden methods:

core/post/tests/test_viewsets.py

```
...
def test_create_anonymous(self, client):
        data = {
            "body": "Test Post Body",
            "author": "test_user"
        }
```

```
        response = client.post(self.endpoint, data)
        assert response.status_code ==
          status.HTTP_401_UNAUTHORIZED


    def test_update_anonymous(self, client, post):
        data = {
            "body": "Test Post Body",
            "author": "test_user"
        }
        response = client.put(self.endpoint +
          str(post.public_id) + "/", data)
        assert response.status_code ==
          status.HTTP_401_UNAUTHORIZED

    def test_delete_anonymous(self, client, post):
        response = client.delete(self.endpoint +
          str(post.public_id) + "/")
        assert response.status_code ==
          status.HTTP_401_UNAUTHORIZED
```

Run the tests again. Great! We've just written tests for the post viewset. You should now have a better understanding of testing with viewsets.

Let's quickly write tests for `CommentViewSet`.

Writing tests for CommentViewSet

Before starting to write the viewset tests, let's also quickly refactor the code for writing the tests. Inside the `core/comment` directory, create a Python package called `tests`. Once it's done, rename the `tests.py` file in the `core/post` directory to `test_models.py` and move it to the `core/comment/tests/` directory.

Inside the same directory, create a new file called `test_viewsets.py`. This file will contain tests for `CommentViewSet`.

Just like in `PostViewSet`, we have two types of users, and we want to write test cases for each of their permissions.

However, before creating comments, we need to add comment fixtures. Inside the `core/fixtures` directory, create a new file called `comment.py` and add the following content:

core/fixtures/comment.py

```
import pytest

from core.fixtures.user import user
from core.fixtures.post import post

from core.comment.models import Comment

@pytest.fixture
def comment(db, user, post):
    return Comment.objects.create(author=user, post=post,
                                  body="Test Comment Body")
```

After that, inside `core/comment/tests/test_viewsets.py`, add the following content first:

core/comment/tests/test_viewsets.py

```
from rest_framework import status

from core.fixtures.user import user
from core.fixtures.post import post
from core.fixtures.comment import comment

class TestCommentViewSet:

    # The comment resource is nested under the post resource

    endpoint = '/api/post/'
```

Next, let's add tests to the list and retrieve comments as authenticated users:

core/comment/tests/test_viewsets.py

```
...
def test_list(self, client, user, post, comment):
        client.force_authenticate(user=user)
        response = client.get(self.endpoint +
```

```
          str(post.public_id) + "/comment/")
        assert response.status_code == status.HTTP_200_OK
        assert response.data["count"] == 1

    def test_retrieve(self, client, user, post, comment):
        client.force_authenticate(user=user)
        response = client.get(self.endpoint +
                              str(post.public_id) +
                              "/comment/" +
                              str(comment.public_id) + "/")
        assert response.status_code == status.HTTP_200_OK
        assert response.data['id'] == comment.public_id.hex
        assert response.data['body'] == comment.body
        assert response.data['author']['id'] ==
          comment.author.public_id.hex
```

Make sure that these tests pass by running the `pytest` command. The next step is to add tests for comment creation, updating, and deletion:

core/comment/tests/test_viewsets.py

```
...
    def test_create(self, client, user, post):
        client.force_authenticate(user=user)
        data = {
            "body": "Test Comment Body",
            "author": user.public_id.hex,
            "post": post.public_id.hex
        }
        response = client.post(self.endpoint +
          str(post.public_id) + "/comment/", data)
        assert response.status_code ==
          status.HTTP_201_CREATED
        assert response.data['body'] == data['body']
        assert response.data['author']['id'] ==
          user.public_id.hex
```

```
    def test_update(self, client, user, post, comment):
        client.force_authenticate(user=user)
        data = {
            "body": "Test Comment Body Updated",
            "author": user.public_id.hex,
            "post": post.public_id.hex
        }
        response = client.put(self.endpoint +
                              str(post.public_id) +
                              "/comment/" +
                              str(comment.public_id) +
                              "/", data)
        assert response.status_code == status.HTTP_200_OK
        assert response.data['body'] == data['body']

    def test_delete(self, client, user, post, comment):
        client.force_authenticate(user=user)
        response = client.delete(self.endpoint +
          str(post.public_id) + "/comment/" +
          str(comment.public_id) + "/")
        assert response.status_code ==
          status.HTTP_204_NO_CONTENT
```

Run the tests again to make sure everything is green. Let's write tests for the anonymous users now.

First of all, we need to make sure that they can access the resources with the GET method:

core/comment/tests/test_viewsets.py

```
...
    def test_list_anonymous(self, client, post, comment):
        response = client.get(self.endpoint +
                              str(post.public_id) +
                              "/comment/")
        assert response.status_code == status.HTTP_200_OK
        assert response.data["count"] == 1
```

```
def test_retrieve_anonymous(self, client, post,
  comment):
    response = client.get(self.endpoint +
      str(post.public_id) + "/comment/" +
      str(comment.public_id) + "/")
    assert response.status_code == status.HTTP_200_OK
```

Next, we need to make sure that an anonymous user can't create, update, or delete a comment:

core/comment/tests/test_viewsets.py

```
def test_create_anonymous(self, client, post):
    data = {}

    response = client.post(self.endpoint +
      str(post.public_id) + "/comment/", data)
    assert response.status_code ==
      status.HTTP_401_UNAUTHORIZED

def test_update_anonymous(self, client, post, comment):
    data = {}

    response = client.put(self.endpoint +
      str(post.public_id) + "/comment/" +
      str(comment.public_id) + "/", data)
    assert response.status_code ==
      status.HTTP_401_UNAUTHORIZED

def test_delete_anonymous(self, client, post, comment):
    response = client.delete(self.endpoint +
      str(post.public_id) + "/comment/" +
      str(comment.public_id) + "/")
    assert response.status_code ==
      status.HTTP_401_UNAUTHORIZED
```

In the preceding cases, the data `dict` is empty because we are expecting error statuses.

Run the tests again to make sure that everything is green!

And voilà. We've just written tests for `CommentViewSet`. We also need to write tests for the `UserViewSet` class, but this will be a small project for you.

Writing tests for the UserViewSet class

In this section, let's do a quick hands-on exercise. You'll write the code for the `UserViewSet` class. It's quite similar to the other tests we've written for `PostViewSet` and `CommentViewSet`. I have provided you with the structure of the class, and all you have to do is to write the testing logic in the methods. The following is the structure you need to build on:

core/user/tests/test_viewsets.py

```
from rest_framework import status

from core.fixtures.user import user
from core.fixtures.post import post

class TestUserViewSet:

    endpoint = '/api/user/'

    def test_list(self, client, user):
        pass

    def test_retrieve(self, client, user):
        pass

    def test_create(self, client, user):
        pass

    def test_update(self, client, user):
        pass
```

Here are the requirements concerning the tests:

- `test_list`: An authenticated user should enable a list of all users
- `test_retrieve`: An authenticated user can retrieve resources concerning a user
- `test_create`: Users cannot create users directly with a `POST` request
- `test_update`: An authenticated user can update a `user` object with a `PATCH` request

You can find the solution to this exercise here: `https://github.com/PacktPublishing/Full-stack-Django-and-React/blob/main/core/user/tests/test_viewsets.py`.

Summary

In this chapter, we learned about testing, the different types of testing, and their advantages. We also introduced testing in Django using Pytest and wrote tests for the models and viewsets. These skills acquired in writing tests using the TDD method help you better design your code, prevent bugs tied to code architecture, and improve the quality of the software. Not to forget, they also give you a competitive advantage in the job market.

This is the last chapter of *Part 1, Technical Background*. The next part will be dedicated to React and connecting the frontend to the REST API we've just built. In the next chapter, we'll learn more about frontend development and React, and we'll also create a React project and run it.

Questions

1. What is testing?
2. What is a unit test?
3. What is the testing pyramid?
4. What is Pytest?
5. What is a Pytest fixture?

Part 2: Building a Reactive UI with React

In *Part 1* of our book, we built the backend of the Postagram application using Django, with authentication features and post and comment management. In this part of the book, you will build a React application representing the UI interface of Postagram, where users will see posts and comments and be able to like posts or comments, upload profile pictures, and visit other users' profiles. At the end of this part, you will have the required knowledge to use React to handle authentication from the frontend side, build UI components from scratch, work with React Hooks such as `useState`, `useContext`, and `useMemo`, and send requests to a REST API and handle the responses.

This section comprises the following chapters:

- *Chapter 6, Creating a Project with React*
- *Chapter 7, Building Login and Registration Forms*
- *Chapter 8, Social Media Posts*
- *Chapter 9, Post Comments*
- *Chapter 10, User Profiles*
- *Chapter 11, Effective UI Testing for React Components*

6

Creating a Project with React

In this chapter, we'll focus on understanding frontend development and creating a web frontend project with React. In previous chapters, we mostly focused on Django and Django Rest. In this chapter, we'll explain the basics of frontend development. Next, we will introduce the React library and create a starting project for the following chapters. Finally, we will learn how to configure our project.

In this chapter, we will cover the following topics:

- Understanding frontend development
- Creating the React project
- Configuring the project
- Useful ES6 and React features

Technical requirements

In this book, we use the Linux OS, but you can find the tools needed for this project on other OSs as well. We'll see how to install Node.js and **Visual Studio Code** (**VS Code**) on your machine in this chapter.

The following GitHub link will also be required: `https://github.com/PacktPublishing/Full-stack-Django-and-React/tree/chap6`.

Understanding frontend development

Frontend development is the part of software development that focuses on the **User Interface** (**UI**). In web development, frontend development is the practice of producing **HTML**, **CSS**, and **JavaScript** for a website or web application.

HTML stands for **HyperText Markup Language**. HTML displays content on the page, such as text, buttons, links, headings, or lists.

CSS is defined as **Cascade Style Sheets**. CSS is used to style the web page. It deals with things such as colors, layouts, and animation. It also helps with the accessibility of your websites so that everyone can easily use your website.

Finally, **JavaScript** is a client-side language that facilitates user interaction and makes dynamic pages. It can help with complex animations, form validation, data fetching from a server, and data submission to the server.

However, as with languages such as Python, while building a frontend application from scratch with HTML, CSS, and JavaScript is definitely possible, it is quite difficult. It requires good knowledge of code architecture and component reusability. In the end, you'll end up creating your own development framework.

But why not directly use some pre-built CSS or JavaScript framework?

Tools such as Vue, Angular, or React can help you write frontend applications with seed and in a smoother way.

In this book, we'll be using React's open source JavaScript library. Let's learn more about React as a library.

What is React?

React is a library that helps developers build reactive UIs as a tree of small reusable pieces called components.

In frontend development, a component is a mixture of HTML, CSS, and JavaScript that captures the logic required to render a small section or a larger UI. Let's analyze the following HTML form to better understand components in frontend development:

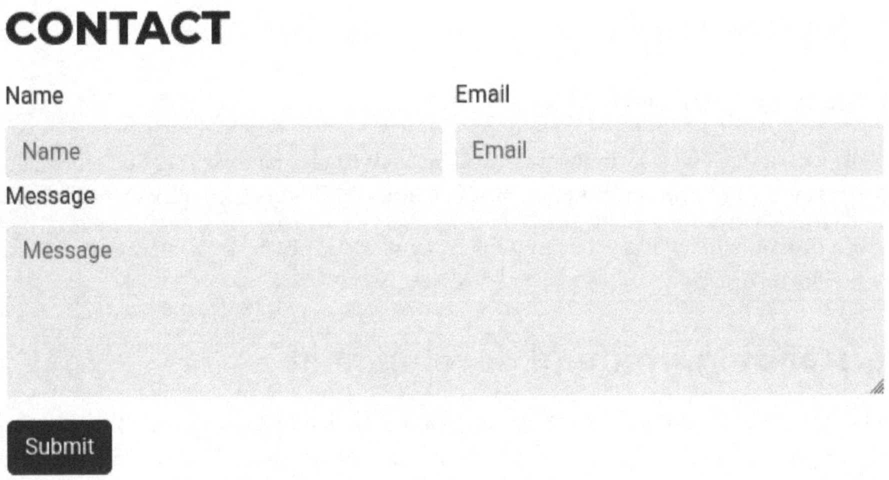

Figure 6.1 – HTML form

As you can see in *Figure 6.1*, in the form, we have defined four components:

- The Name input
- The Email input
- The Message input
- The **Submit** button

Each of these components has its own logic. For example, the **Submit** button will validate the form and save or send the message to a remote source.

React is defined as a library instead of a framework because it only deals with UI rendering and leaves many of the other things that are important in development to the developers or other tools.

To build a React application, you'll need the following stack:

- **Application code**: React, Redux, ESLint, Prettier, and React Router
- **Build tools**: Webpack, Uglify, `npm/yarn/pnpm`, and `babel`
- **Testing tools**: Jest and Enzyme

You'll need to add these dependencies to your React project to optimize and perform some tasks – that's where React differs from tools such as Angular, which comes with its own stack for routing, for example.

Now that we better understand React, let's create a React project.

Creating the React project

Before creating the React project, we need to have tools installed for a better development experience. These tools are drivers, editors, and plugins basically. Let's start by installing Node.js.

Installing Node.js

Node.js is an open source and powerful JavaScript-based server-side environment. It allows developers to run JavaScript programs on the server side, even though JavaScript is natively a client-side language.

Node.js is available for multiple OSs, such as Windows, macOS, and Linux. In this book, we are working on a Linux machine and Node.js should normally be installed already by default.

For other OSs, you can find the installation package at `https://nodejs.org/en/download/`. Download the latest **Long-Term Support** (**LTS**) version for your OS.

When visiting the link, you'll have an output similar to the following screenshot:

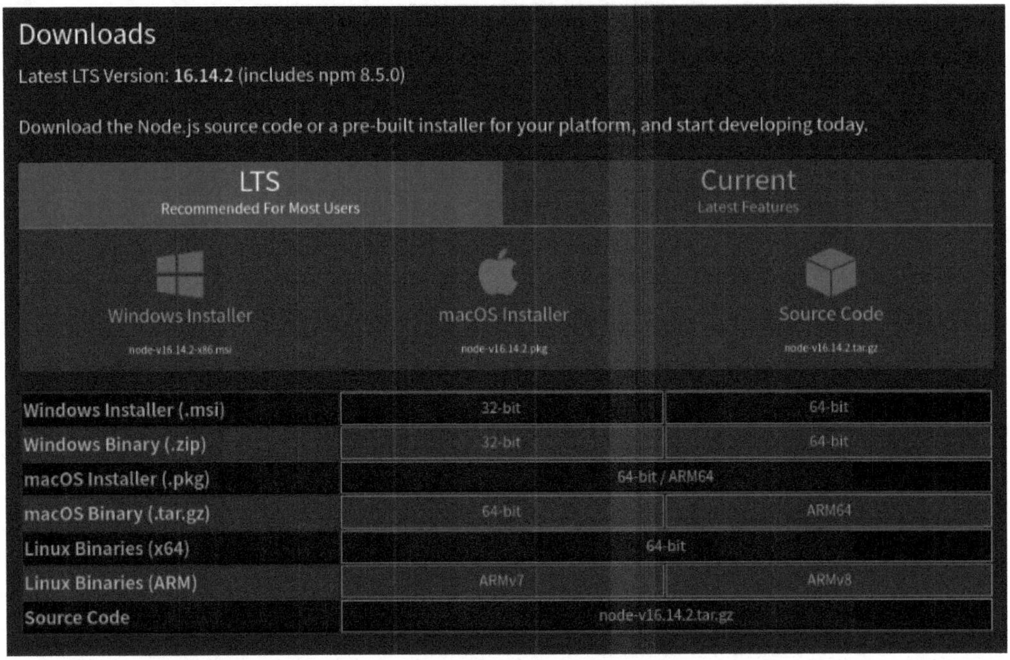

Figure 6.2 – Node.js installers

To check whether Node.js has been installed on your Linux machine, open the Terminal and enter the following commands:

```
node -v
yarn -v
```

These commands should show you the versions of Node.js and `yarn` installed:

Figure 6.3 – node and yarn versions

If you don't have `yarn` installed on your machine, you can install it with the `npm` package:

```
npm install -g yarn
```

`yarn` and `npm` are package managers for JavaScript. We'll use the `yarn` package manager a lot in upcoming chapters to install packages, run tests, or build a production-ready version of the frontend. However, feel free to use `npm` if you want. Just don't forget that the commands are slightly different.

The basic tools to develop with JavaScript have now been installed. Next, we will need to install VS Code and configure it to make JavaScript development easier.

Installing VS Code

VS Code is an open source code editor developed and maintained by Microsoft. It supports multiple programming languages and with the plugins and extensions, you can easily transform it into a powerful IDE. However, you can also use other editors such as **Atom**, **Brackets**, or the powerful **IDE WebStorm**. Feel free to use what you are familiar with.

VS Code is available for Windows, macOS, and Linux and you can download the right version for your OS at `https://code.visualstudio.com/`.

Once it's installed and opened, you'll see the following window:

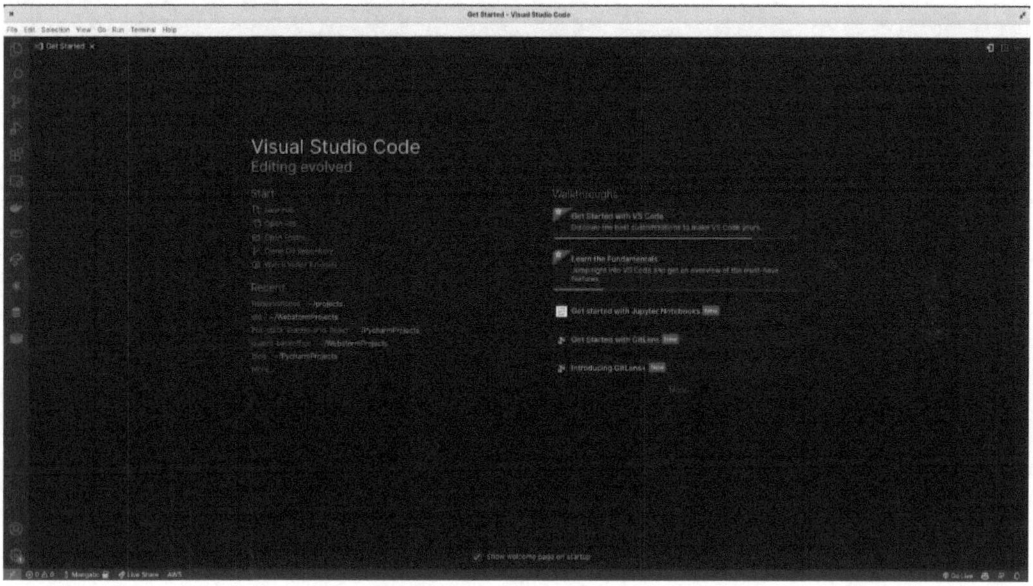

Figure 6.4 – VS Code window

VS Code comes with an integrated terminal that you can use to create and run React apps. Note also that you can open projects with VS Code using the following command in the terminal in the directory of the project:

```
code .
```

You can find the integrated terminal in the **View | Integrated Terminal** menu.

With the basics of VS Code explored, let's add the needed extensions to make React development more enjoyable.

Adding VS Code extensions

Every programming language and framework comes with a lot of extensions available to make development easier and more enjoyable. These extensions include code snippets, testing, project environment configuration, and code formatting. In VS Code, if you open **Extensions** in the activity bar (the bar on the left), you can find a search bar to look for different extensions.

For the React project, let's start by adding the **ES7+ React/Redux/React-Native/JS snippets** extension. This extension will suggest code snippets when writing code in React files. It should look something like this:

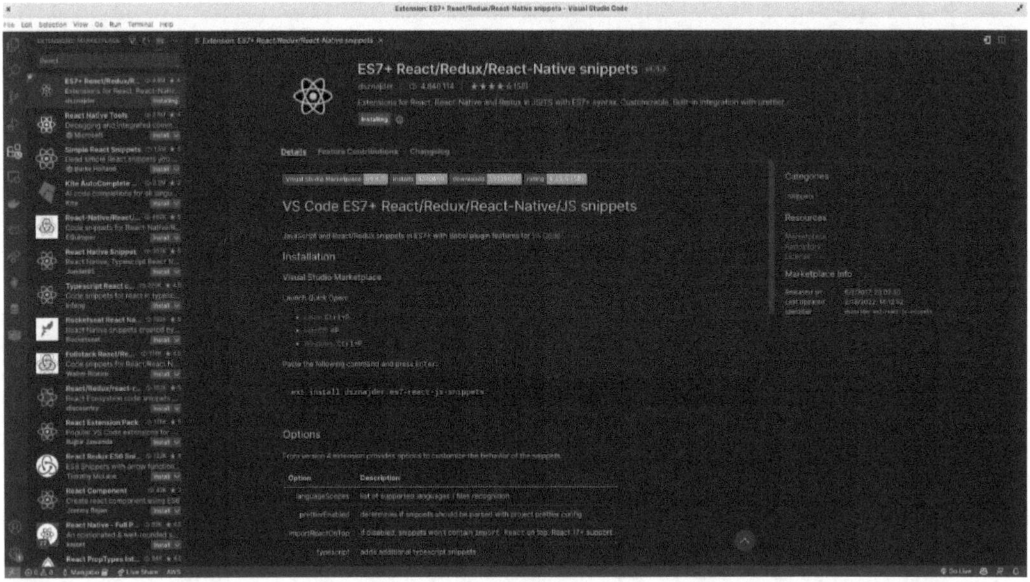

Figure 6.5 – ES7 + React/Redux/React-Native/JS snippets extension

After that, let's install the **ESLint** extension. It'll help you find typos and syntax errors quickly by automatically formatting the code and showing formatting errors. This makes the ES code formatting rules easy to understand. The ESLint extension looks like this:

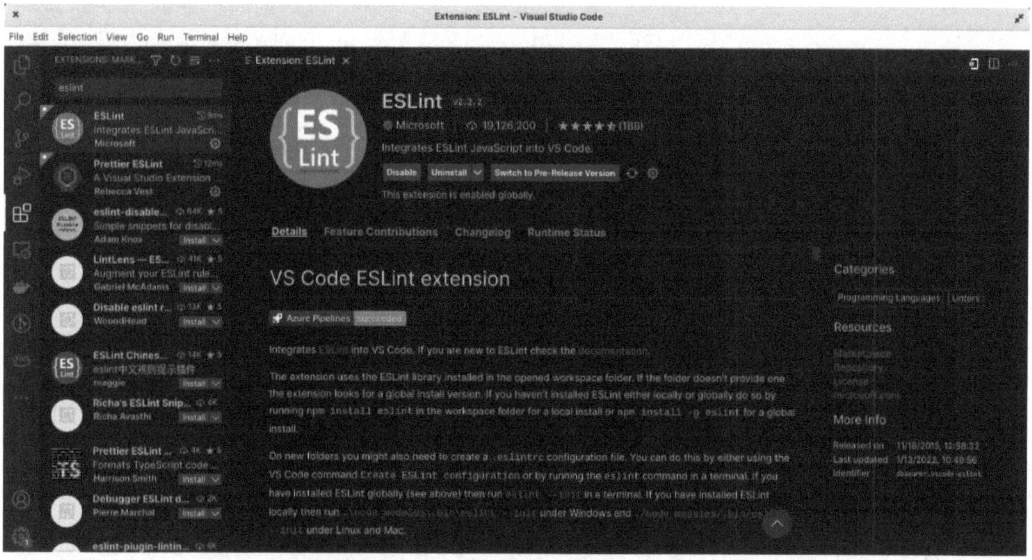

Figure 6.6 – The ESLint extension

Next, we will add another VS Code extension called **Prettier**. Prettier is a code formatter that not only makes your code visually appealing but also much more structured for readability. You can find a VS Code extension that can help you format your code automatically after saving your code. The extension looks something like this:

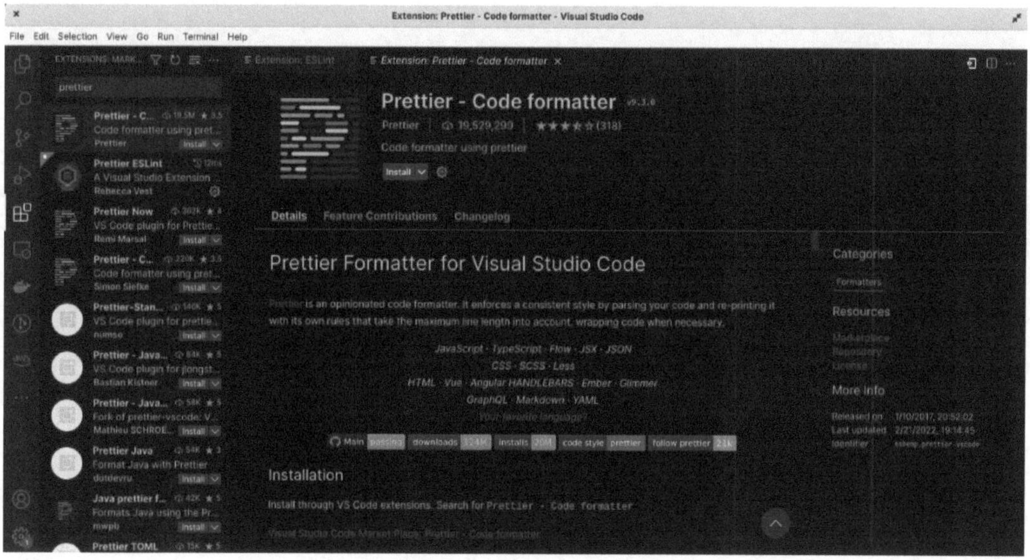

Figure 6.7 – Prettier code formatter

And finally, but optionally, we have **indent-rainbow**. If you have many blocks of code with parents and children, it can become quite difficult to read. This extension will make JavaScript code with indentation more readable. It looks like this:

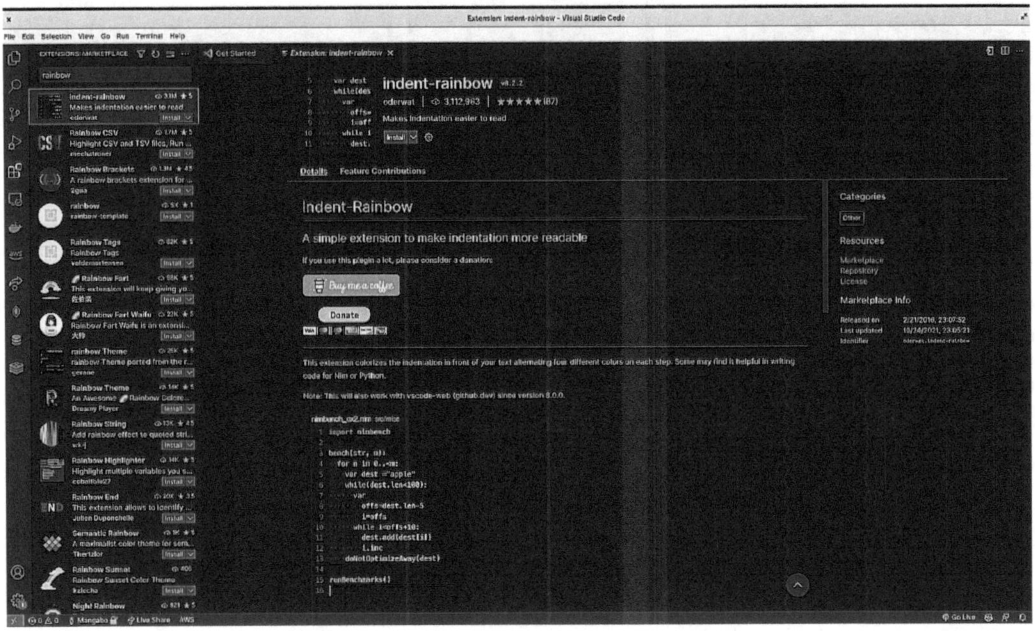

Figure 6.8 – The indent-rainbow extension

Great! With these extensions installed in VS Code, we can now move on to creating the React application.

Creating and running a React app

With Node.js and VS Code installed and configured, we have everything we need to create our first React.js application.

To create our React app, we'll be using `create-react-app` (`https://github.com/facebook/create-react-app`), a simple command for creating a modern web React application. Follow these steps to create your first React application and modify the code:

1. Run the following command to create a React application:

```
yarn create react-app social-media-app
```

This command will create a React application named `social-media-app`. If you are using `npm`, then replace `yarn` with `npx`. After installation, you will have an output similar to the following screenshot:

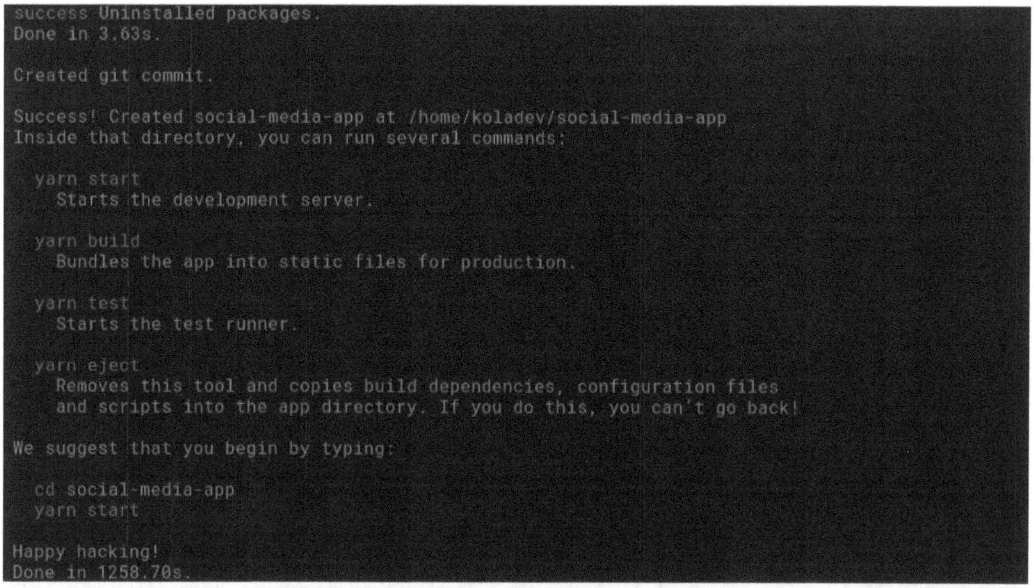

Figure 6.9 – The React project creation terminal output

Inside `social-media-app`, you'll find a file called `package.json`. This file contains all the configurations for the JavaScript project, starting from basic information about the project, such as the name, the version, and the developers, but it also includes a list of installed packages and the scripts related to starting the server, for example.

2. Run the created React application with the following command:

```
yarn start
```

3. Open your browser and specify `localhost:3000` as your web link. Once done, it will look something like this:

Figure 6.10 – Running the React application

The application is running. Now, let's modify the code in the React application.

4. Open the `App.js` file from the `src` folder in the VS Code editor.
5. Modify the text inside the `App.js` file from `Learn React` to `Hello World` and save the file:

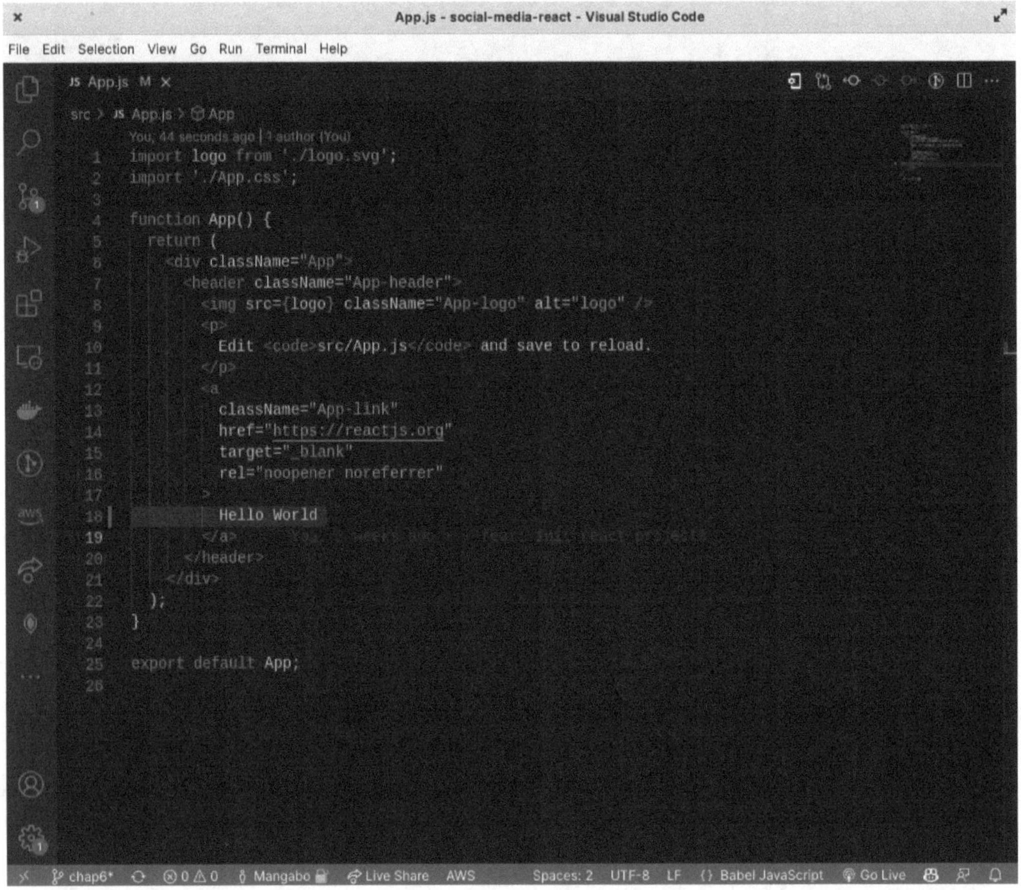

Figure 6.11 – The App.js code

6. Check the browser again and you'll see the changes:

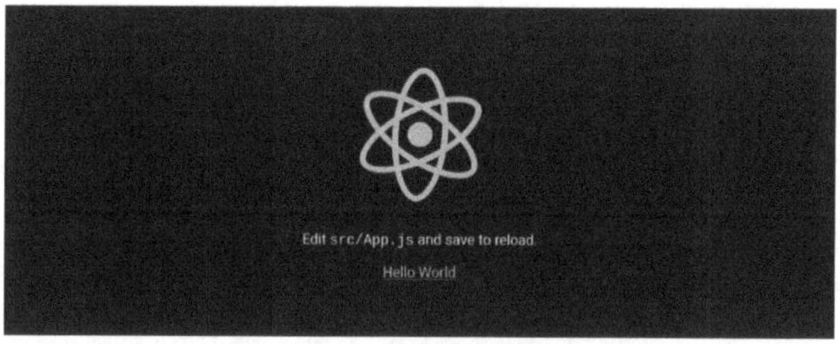

Figure 6.12 – Modified React application

React has a hot reload feature, meaning that any changes made to a file in the project are reflected in the rendering of the web application.

Great! We've just installed a React application and modified the code.

Let's install some tools in the browser for debugging the React application.

Installing a debugging plugin in the browser

To debug React applications, we have to install React Developer Tools, a plugin available on Chrome, Firefox, and Edge browsers. You can find the plugin for the Chrome version at `https://chrome.google.com/webstore/category/extensions` and the Firefox version at `https://addons.mozilla.org`. The plugin looks something like this:

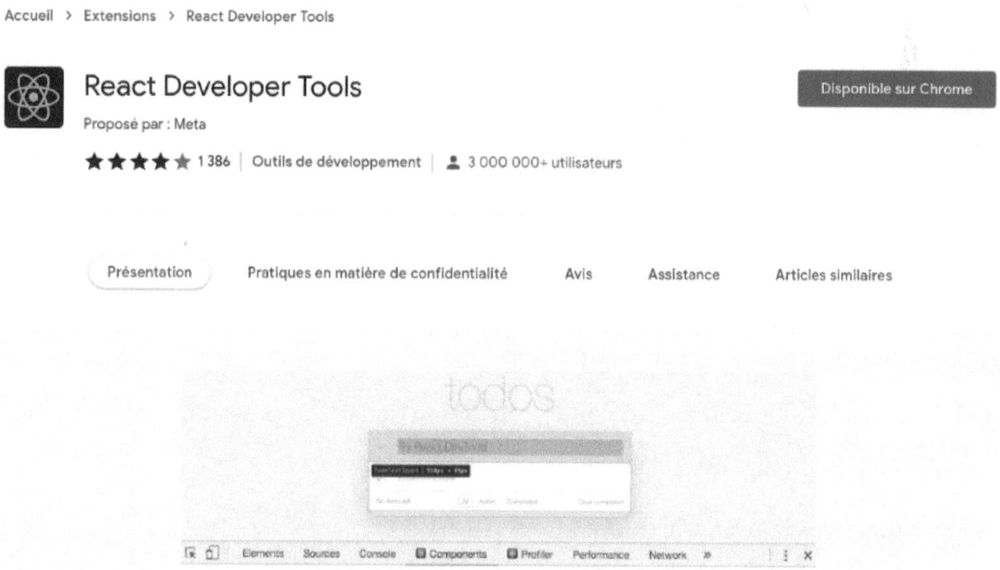

Figure 6.13 – The React browser plugin

Once it's installed, you can open the developer tools by pressing *Ctrl* + *Shift* + *I* (or *F12*) in the Chrome browser. The following screenshot shows the developer tools in the Firefox browser:

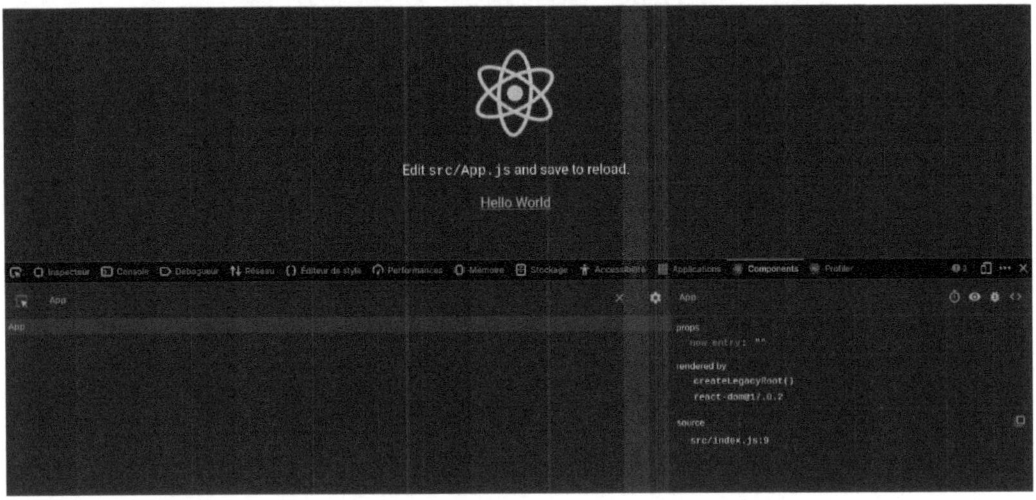

Figure 6.14 – React application with the open React extension

This tool will be useful for finding bugs and debugging the application in the development phase.

The project is created and can now be successfully run. Let's install and configure some packages for routing and styling in the next section.

Configuring the project

Before starting to write the authentication flow, let's make sure that the project is ready for coding. In this section, we will configure styling and routing, and allow the request on the API.

Let's start with routing first.

Adding React Router

Routing in a frontend application represents everything that deals with moving from one view to another and loading the right page using the right URL.

React doesn't come with an integrated routing package, so we'll use the `react-router` package.

You can install the package using the following command:

```
yarn add react-router-dom@6
```

Then, edit the `index.js` file like so:

src/index.js

```
import React from "react";
import ReactDOM from "react-dom/client";
import { BrowserRouter } from "react-router-dom";
import "./index.css";
import App from "./App";

// Creating the root application component
const root = ReactDOM.createRoot(document.
getElementById("root"));

root.render(
  <React.StrictMode>
    <BrowserRouter>
      <App />
    </BrowserRouter>
  </React.StrictMode>
);
```

In the preceding code block, we imported the `BrowserRouter` component and wrapped it inside the `React.StrictMode` component which helps us receive warnings in the development mode (`https://reactjs.org/docs/strict-mode.html`), and finally, the `App` component is wrapped inside the `BrowserRouter` component.

With React Router configured, we can freely move on to installing React Bootstrap for styling.

Adding React Bootstrap

React is easily configurable with CSS frameworks. For this project, for the sake of simplicity and rapidity of development, we'll go with Bootstrap.

Fortunately, the React ecosystem provides a package called `react-bootstrap` independent of JQuery.

Run the following command to install the package:

```
yarn add react-bootstrap bootstrap
```

Next, import the bootstrap CSS file into the index.js file like so:

src/index.js

```
...
import 'bootstrap/dist/css/bootstrap.min.css';
import App from "./App";
...
```

With react-router and react-bootstrap installed, let's create a quick page using both of these in the next subsection.

Creating the Home page

Creating a page in React using React Router follows this pattern most of the time:

- Creating the component and the page
- Registering the page in BrowserRouter with an URL

Follow these steps to create the Home page:

1. Create a directory in src called pages.
2. Inside the pages directory, create a new file called Home.jsx:

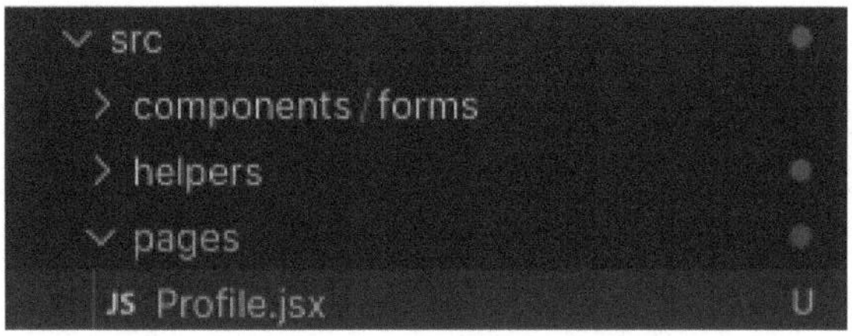

Figure 6.15 – The pages folder structure

This file will contain the UI for the Profile page.

3. Add the following text to the Home.jsx file to ensure that authentication is working properly:

src/pages/Home.jsx

```
import React from "react";

function Home() {
  return (
```

```
    <div>
      <h1>Profile</h1>
      <p>
        Welcome!
      </p>
    </div>
  );
}

export default Home;
```

4. Register this page in the `App.js` file:

src/App.js

```
import React from "react";
import { Route, Routes } from "react-router-dom";
import Home from "./pages/Home";


function App() {
  return (
    <Routes>
      <Route path="/" element={<Home />} />
    </Routes>
  );
}

export default App;
```

To register a page with React Router, you use the `<Route />` component and pass two props:

- The path
- The component element

5. With the preceding code added, make sure that the React project is running. You can check the page at `http://127.0.0.1:3000`:

Figure 6.16 – Home page

Great! With this added, let's quickly configure the Django project to avoid some request issues in the next section.

Configuring CORS

CORS stands for **cross-origin resource sharing**. It's a browser mechanism that enables controlled access to resources located outside of a given domain.

It helps prevent cross-domain attacks or unwanted requests. In the case of this project, the React project is running on `http://127.0.0.1:3000`.

If we try to make some requests from the browser, we'll receive an error. Open the React application at `http://127.0.0.1:3000` and open **Developer Tools**:

Figure 6.17 – Opening the developer tools

Also, make sure that the Django server is running.

In the console, enter the following line:

```
fetch("http://127.0.0.1:8000/api/post/")
```

You'll receive an error:

Figure 6.18 – A CORS error when making a request

Let's quickly configure the Django API side by following these steps:

1. Enable **CORS** with Django REST by using `django-cors-headers`:

```
pip install django-cors-headers
```

2. If the installation of the `django-cors-headers` package is complete, go to your `settings.py` file and add the package into `INSTALLED_APPS` and the middleware:

CoreRoot/settings.py

```
INSTALLED_APPS = [
    ...
    'corsheaders',
    ...
]

MIDDLEWARE = [
    ...
    'corsheaders.middleware.CorsMiddleware',
    'django.middleware.common.CommonMiddleware',
    ...
]
```

3. Add these lines at the end of the `settings.py` file:

CoreRoot/settings.py

```
CORS_ALLOWED_ORIGINS = [
    "http://localhost:3000",
    "http://127.0.0.1:3000"
]
```

4. Make the request again in **Developer Tools**.

   You will see that the request has passed, and we are good now. The API is ready to accept requests from the React application:

Figure 6.19 – Trying a successful request in Developer Tools

With the React project configured with the backend for a better development experience, we can now explore the **ES6** (**ECMAScript 6**) and React features that we will use a lot in upcoming chapters.

Useful ES6 and React features

JavaScript and React are evolving languages and technologies, incorporating exciting, new features each year. **ES6**, also known as **ECMAScript 2015**, is a significant enhancement in the JavaScript language that allows developers to write programs for complex applications with better techniques and patterns.

With React, we have moved from writing classes to writing components using functions and React Hooks. In this section, we will quickly explore the ES6 syntaxes, React concepts, and React Hooks that we will use in the following chapters.

const and let

The `const` keyword was introduced in **ES6** and it makes any variables immutable when declared with the keyword. When using the `const` keyword, variables can't be redeclared nor reassigned. Here's an example of its usage:

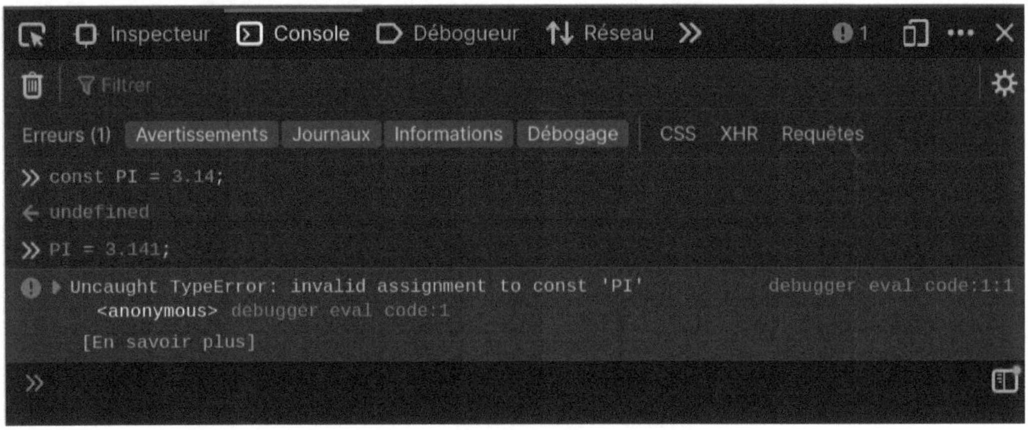

Figure 6.20 – Usage of the const keyword

On the other hand, `let` is used to declare a variable that can be reassigned to a new value. This is useful when you want to create a variable that can change over time, such as a counter or an iterator. Here's an example:

```
let counter = 0;
// This is allowed because counter is not a constant
counter++;
```

In general, it is a good practice to use `const` by default, and only use `let` when you need to reassign a variable. This can help to make your code more readable and prevent accidental reassignments.

Now that we understand the usage of `const` and `let` keywords, let's move on to understanding template literals in JavaScript.

Template literals

In JavaScript, template literals are a way to define string values that can contain placeholders for dynamic values. They are represented by the backtick (`) character and use the dollar sign (`$`) and curly braces ({ }) to insert expressions into the string.

Here is an example of a template literal:

```
const name = 'World;
const message = `Hello, ${name}!`;
console.log(message); // "Hello, World!"
```

In this example, we defined a template literal named `message` that contains a placeholder for the `name` variable. When the template literal is evaluated, the `name` variable is inserted into the string and the resulting string is logged to the console.

Template literals provide a more convenient and readable way to create strings with dynamic values compared to using the traditional string concatenation operator (+). They also support **string interpolation**, which means that you can insert expressions into the string, as well as multiline strings. Here's an example:

```
const a = 10;
const b = 20;
const message = `The sum of ${a} and ${b} is ${a + b}.`;
console.log(message); // "The sum of 10 and 20 is 30."
```

In the preceding example, we defined a template literal called `message` that contains multiple expressions that are inserted into the string when the template literal is evaluated. This allows us to create a string with dynamic values that is more readable and concise than when using string concatenation.

Now that we understand what template literals are, let's explain JSX styling in React.

JSX styling

JSX is a syntax extension to JavaScript that allows you to write JavaScript code that looks like HTML. It was introduced by Facebook as part of the React library, but it can be used with other JavaScript libraries and frameworks as well. Here is an example of how you might use JSX in a React component:

```
import React from 'react';

function Component() {
  return (
    <div>
      <h1>Hello, world!</h1>
      <p>This is some text.</p>
    </div>
  );
}
```

In the preceding example, we defined a React component called `Component` that returns some JSX code. The JSX code looks like HTML, but it is transformed into JavaScript by the React library, which generates the appropriate elements and attributes in the DOM.

When you write JSX, you can use JavaScript expressions inside the curly braces ({ }) to insert dynamic values into the JSX code. This allows you to easily create dynamic and interactive UIs using JSX:

```
import React from 'react';

function Component({ name }) {
  return (
    <div>
      <h1>Hello, {name}!</h1>
      <p>This is some text.</p>
    </div>
  );
}
```

In the preceding example, we defined a React component called `Component` that takes a name prop and inserts it into the JSX code using a JavaScript expression. This allows us to create a dynamic and personalized greeting for the user.

Now that we understand how JSX works with React, let's explain the concept of props and states.

Props versus states

In React, props and states are two different ways to manage data in a component.

Props are short for **properties** and are used to pass data from a parent component to a child component. Props are read-only, which means that a child component cannot modify the props passed to it by the parent component:

```
import React from 'react';

function ParentComponent() {
  return (
    <ChildComponent
      name="John Doe"
      age={25}
    />
  );
}

function ChildComponent({ name, age }) {
  return (
    <p>
      My name is {name} and I am {age} years old.
    </p>
  );
}
```

In the preceding code, we defined a parent component called `ParentComponent` that renders a child component called `ChildComponent` and passes two props to the child component (`name` and `age`). The child component receives these props as arguments and uses them to render the content of the component. Because props are read-only, the child component cannot modify the `name` and `age` props passed to it by the parent.

On the other hand, a state is a way to manage data in a component that can be modified by the component itself. The state is private to the component and can only be modified using special React methods, such as `useState`.

Here is an example of how you might modify a state in a React component:

```
import React, { useState } from 'react';

function Counter() {
  // Use useState to manage the state of the counter
  const [count, setCount] = useState(0);

  // Function to increment the counter
  function handleIncrement() {
    setCount(count + 1);
  }

  return (
    <div>
      <p>The counter is at {count}.</p>
      <button onClick={handleIncrement}>Increment</button>
    </div>
  );
}
```

In the preceding code, we define a component called `Counter` that uses the `useState` Hook to manage the state of a counter. The `useState` Hook returns an array with two elements, the current value of the state (in this case, `count`) and a function to update the state (in this case, `setCount`).

In the render method of the component, we display the value of the count state and define a button that, when clicked, calls the `handleIncrement` function to update the count state. This causes the component to re-render and display the updated value of the count state.

Now that we understand the difference between props and state better, let's dive deeper into understanding the `useState` Hook.

> **Important note**
>
> `useState` is a Hook in React that allows you to add a state to functional components. In other words, `useState` allows you to manage the state of your component, which is an object that holds information about your component and can be used to re-render the component when this state changes.

The Context API

The **Context API** is a way to share data between different components in a React application. It allows you to pass data through the component tree without having to pass props down manually at every level. Here is an example of how you might use the Context API in a React application:

```
// Create a context for sharing data
const Context = React.createContext();

function App() {
  // Set some initial data in the context
  const data = {
    message: 'Hello, world!'
  };

  return (
    // Provide the data to the components inside the
    // Provider
    <Context.Provider value={data}>
      <Component />
    </Context.Provider>
  );
}

function Component() {
  // Use the useContext Hook to access the data in the
  // context
  const context = React.useContext(Context);

  return (
    <p>{context.message}</p>
  );
}
```

In the preceding code, we use the `React.createContext` method to create a new context object, which we call `Context`. We then provide some initial data to the context by wrapping our top-level component in a `Context.Provider` component and passing the data as the value prop. Finally, we use the `useContext` Hook in `Component` to access the data in the context and display it in the component.

There is also another Hook that we will use in this book. Let's explain the `useMemo` Hook in the next section.

useMemo

`useMemo` is a Hook in React that allows you to optimize the performance of your components by memoizing expensive calculations. It works by returning a memoized value that is only recalculated if one of the inputs to the calculation changes.

> **Important note**
>
> **Memoization** is a technique used in computer programming to speed up programs by storing the results of expensive function calls and returning the cached result when the same inputs are given again. This can be useful for optimizing programs that make many repeated calculations with the same input.

Here is an example of how you might use `useMemo` to optimize the performance of a component:

```
import React, { useMemo } from 'react';

function Component({ data }) {
  // Use useMemo to memoize the expensive calculation
  const processedData = useMemo(() => {
    // Do some expensive calculation with the data
    return expensiveCalculation(data);
  }, [data]);


  return (
    <div>
      {/* Use the processed data in the component */}
      <p>{processedData.message}</p>
    </div>
  );
}
```

In the preceding code, we use `useMemo` to memoize the result of an expensive calculation that we do with the data prop passed to the component. Because `useMemo` only recalculates the value if the data prop changes, we can avoid making the expensive calculation every time the component is re-rendered, which can improve the performance of our application.

In the React project that we will build in the next chapter, we will work with forms using React and the Hooks provided by the React library. Let's learn more about controlled and uncontrolled components.

Handling forms – controlled components and uncontrolled components

A controlled component is a component in React that is controlled by the state of the parent component. This means that the value of the input field is determined by the value prop passed to the component, and any changes to the input are handled by the parent component.

Here is an example of a controlled component:

```
import React, { useState } from 'react';

function Form() {
  // Use useState to manage the state of the input field
  const [inputValue, setInputValue] = useState('');

  // Function to handle changes to the input field
  function handleChange(event) {
    setInputValue(event.target.value);
  }

  return (
    <form>
      <label htmlFor="name">Name:</label>
      <input
        type="text"
        id="name"
        value={inputValue}
        onChange={handleChange}
      />
    </form>
  );
}
```

In the preceding code, we use `useState` to manage the state of the input field and the `handleChange` function to update the state when the input is changed. Because the value of the input is determined by the `inputValue` state variable, the input is considered to be a controlled component.

On the other hand, an uncontrolled component is a component in React that manages its own state internally. This means that the value of the input field is determined by the `defaultValue` prop passed to the component, and any changes to the input are handled by the component itself.

Here is an example of an uncontrolled component:

```
import React from 'react';

function Form() {
  // Use a ref to manage the state of the input field
  const inputRef = React.useRef();

  // Function to handle the form submission
  function handleSubmit(event) {
    event.preventDefault();

    // Do something with the input value here
    // For example, you might send the input value to an
    // API or save it to the database
    sendInputValue(inputRef.current.value);

    // Clear the input after submitting
    inputRef.current.value = '';
  }

  return (
    <form onSubmit={handleSubmit}>
      <label htmlFor="name">Name:</label>
      <input
        type="text"
        id="name"
        defaultValue=""
        ref={inputRef}
      />
    </form>
  );
}
```

In this example, we used `ref` to manage the state of the input field and the `handleSubmit` function to handle the form submission. Because the value of the input is determined by the `defaultValue` prop and managed internally by the component, the input is considered to be an uncontrolled component.

In this section, we have explored most of the React and ES6 features that we will use in the next chapters. We will mostly be working with React Hooks, JSX, and interesting ES6 features such as template literals and `let/const` keywords.

Summary

In this chapter, we've explained frontend development and created a React application by installing **Node.js** and **VS Code**. We then configured it for better development using **VS Code**. React will also run in the browser, so we installed some plugins that will make debugging easier.

Then, we started coding a little bit with basic configuration for routing, styling, and CORS configuration to allow requests on the Django API. Finally, we explored the React and ES6 features that we will be using in the next chapters.

In the next chapter, we'll familiarize ourselves more with React by building a login and register page while explaining component-driven development.

Questions

1. What are Node.js and `yarn`?
2. What is frontend development?
3. How do you install Node.js?
4. What is VS Code?
5. How do you install extensions in VS Code?
6. What is the purpose of hot reloading?
7. How do you create a React application with `create-react-app`?

7
Building Login and Registration Forms

Registration and login are essential features of web applications that have users. Even if an authentication flow can be handled directly with simple requests, there is also a need to have logic working behind the UI to manage the authentication and session, especially if we are using a **JSON web token** (**JWT**).

In this chapter, we'll create login and registration forms with React. There is a lot to do and learn here, but here's what this chapter will cover:

- Configuration of a CSS framework in a React project
- Adding protected and public pages to an application
- Creating a page for registration
- Creating a page for login
- Creating a welcome page after the login or registration is successful

By the end of the chapter, you will be able to build registration and login pages using React, and you will know how to manage JWT authentication from the frontend.

Technical requirements

Make sure to have VS Code installed and configured on your machine.

You can find the code for this chapter at `https://github.com/PacktPublishing/Full-stack-Django-and-React/tree/chap7`.

Understanding the authentication flow

We've already explored authentication on a social media project from a backend perspective in *Chapter 2, Authentication and Authorization Using JWTs*. But how does this manifest in the React application?

Well, things will be a little bit different. To quickly recapitulate, we have a registration and a login endpoint. These endpoints return the user objects with two tokens:

- **An access token with a lifetime of 5 minutes**: This token helps with authenticating on the server side when requesting without the need to log in again. Then, we can access resources and perform actions on these resources.
- **A refresh token**: This token helps you to retrieve another access token if one has already expired.

With this data coming from the server, we can manage authentication from the React application side like so. When a registration or a login is successful, we store the returned response in the client's browser; we'll use `localStorage` for this.

The `localStorage` property helps us to work with the browser storage, enabling browsers to store key-value pairs in the browser. Two methods will be used with `localStorage`: `setItem()` to set a key-value pair and `getItem()` to access the values.

Then, for each request sent to the server, we add the Authorization header to the request containing the access token retrieved from `localStorage`. If the request returns a `401` error, it means that the token has expired. If this happens, we send a request to the refresh endpoint to get a new access token, using the refresh token also retrieved from `localStorage`. And with this access token, we resend the failed request.

If we receive a `401` error again, it means that the refresh token has expired. Then, the user will be sent to the login page to log in again, retrieve new tokens, and store them in `localStorage`.

Now that we understand the authentication flow from the frontend side, let's write the requests service we will use for data fetching and performing CRUD actions.

Writing the requests service

Making requests in JavaScript is relatively easy. The node environment and the browser provide native packages such as `fetch` to allow you to request a server. However, this project will use the `axios` package for HTTP requests.

Axios is a popular library mainly used to send asynchronous HTTP requests to REST endpoints. Axios is the perfect library for CRUD operations. However, we will also install `axios-auth-refresh`. This simple library assists with an automatic refresh of tokens via `axios` interceptors. To install the `axios` and `axios-auth-refresh` packages, follow these steps:

1. In the `social-media-app` directory, add the `axios` and `axios-auth-refresh` packages by running the following command:

   ```
   yarn add axios axios-auth-refresh
   ```

2. Once it's installed, create a directory called `helpers` in the `src` folder of the React project, and once it's done, add a file called `axios.js`:

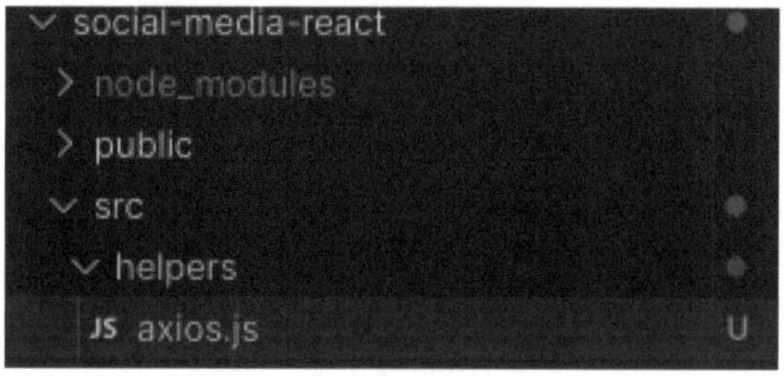

Figure 7.1 – The path of the helper.js file

Now, let's make the import and write the basic configurations, such as the URL and some headers. Take a look at the following code block:

```
import axios from "axios";
import createAuthRefreshInterceptor from "axios-auth-refresh";

const axiosService = axios.create({
 baseURL: "http://localhost:8000",
 headers: {
   "Content-Type": "application/json",
 },
});
```

In the preceding code block, we have added the `Content-Type` header for the `POST` requests. The following figure shows the authentication flow we'll follow in this book:

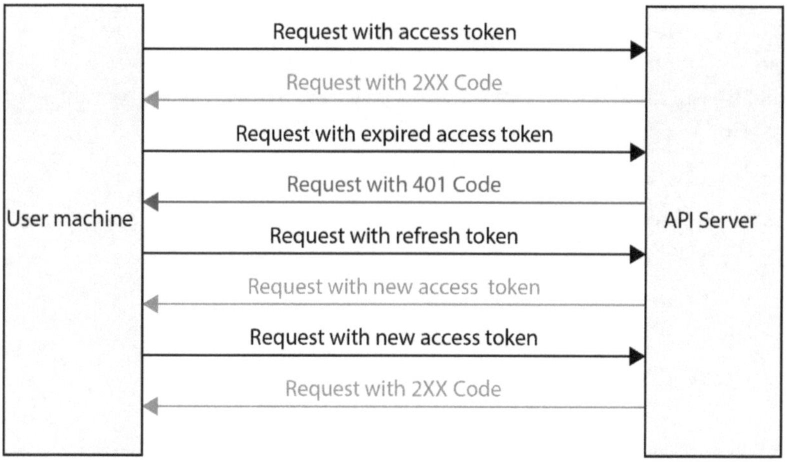

Figure 7.2 – Authentication flow with access/refresh tokens

In the preceding figure, note the following points:

- Every time we are requesting with `axiosService`, we retrieve the access token from `localStorage` and create a new header authorization using the access token
- The access token will expire if the request is made and a `400` status code is returned
- We retrieve the refresh token from `localStorage` and make a request to retrieve a new access token
- Once done, we register the new access token in `localStorage` and restart the previously failed request
- Yet, if the refresh token request has failed too, we simply remove `auth` from `localStorage` and send the user to the login screen

Let's implement the previously described flow in the `axios.js` file by following these steps:

1. First we will write a request interceptor to add headers to the request:

```
axiosService.interceptors.request.use(async (config) => {
  /**
   * Retrieving the access token from the localStorage
     and adding it to the headers of the request
   */
  const { access } =
    JSON.parse(localStorage.getItem("auth"));
  config.headers.Authorization = `Bearer ${access}`;
  return config;
});
```

Note that we can use the object-destructuring syntax to extract property values from an object in JavaScript. In pre-ES2015 code, it probably goes like this:

```
var fruit = {
 name: 'Banana',
 scientificName: 'Musa'
};
var name      = fruit.name;
var scientificName = fruit.scientificName;
```

If you have a lot of properties to extract from an object, it can quickly become long. That's where object destructuring comes in handy:

```
var fruit = {
 name: 'Banana',
 scientificName: 'Musa'
};
var  { name, scientificName } = fruit;
```

You can learn more about the syntax at `https://developer.mozilla.org/en-US/docs/Web/JavaScript/Reference/Operators/Destructuring_assignment`.

2. After that, we will resolve the requests and return a resolved or rejected promise:

```
axiosService.interceptors.response.use(
  (res) => Promise.resolve(res),
  (err) => Promise.reject(err),
);
```

3. This last step is the icing on the cake. Create a function that contains the refresh `auth` logic. This function will be called whenever the failed request returns a `401` error:

```
const refreshAuthLogic = async (failedRequest) => {
 const { refresh } =
   JSON.parse(localStorage.getItem("auth"));
 return axios
   .post("/refresh/token/", null, {
     baseURL: "http://localhost:8000",
     headers: {
       Authorization: `Bearer ${refresh}`,
     },
   })
   .then((resp) => {
     const { access, refresh } = resp.data;
     failedRequest.response.config.headers[
       "Authorization"] = "Bearer " + access;
     localStorage.setItem("auth", JSON.stringify({
                             access, refresh }));
   })
   .catch(() => {
```

```
        localStorage.removeItem("auth");
    });
};
```

4. And finally, initialize the authentication interceptor and create a custom fetcher too:

```
createAuthRefreshInterceptor(axiosService,
refreshAuthLogic);

export function fetcher(url) {
 return axiosService.get(url).then((res) => res.data);
}

export default axiosService;
```

The fetcher will be used to make `GET` requests on the API resources. Great! The fetching logic is implemented, and we can move on to registering a user. But before that, we need to define protected routes in the project.

Protected routes

Routing with the condition on a frontend application is a big plus, as it helps with a better user experience. For example, if you are not logged in to Twitter and want to check a profile or comment, you will be redirected to the login page. These are protected pages or actions, so you must log in before accessing these resources. In this section, we'll write a `ProtectedRoute` component using `React-Router` components.

Creating a protected route wrapper

To create a protected route wrapper, follow these steps:

1. Create a new directory in the `src` directory called `routes`.
2. Inside the newly created directory, create a file called `ProtectedRoute.jsx`.
3. Once the file is created, import the needed libraries:

src/routes/ProtectedRoute.jsx

```
import React from "react";
import { Navigate } from "react-router-dom";
...
```

4. Write the following logic for the protected routes:

```
...
function ProtectedRoute({ children }) {
 const { user } =
   JSON.parse(localStorage.getItem("auth"));

 return auth.account ? <>{children}</> : <Navigate
   to="/login/" />;
}

export default ProtectedRoute;
...
```

In the preceding code snippet, we are retrieving the user property from `localStorage`.

We then use this property to check whether we should redirect the user to the login page or render the page (`children`). If **user** is null or undefined, it means that the user has not logged in, so we redirect the user to the login page, otherwise, we give access to the asked page.

5. Then, inside the `App.js` file, let's rewrite the content:

src/App.js

```
import React from "react";
import {
 Route,
 Routes
} from "react-router-dom";
import ProtectedRoute from "./routes/ProtectedRoute";
import Home from "./pages/Home";

function App() {
 return (
   <Routes>
     <Route path="/" element={
       <ProtectedRoute>
         <Home />
       </ProtectedRoute>
     } />
```

```
            <Route path="/login/" element={<div>Login</div>} />
        </Routes>
    );
}

export default App;
```

Now, the default location will be the profile page. However, with no credentials in the store, the user will be redirected to the login page.

Great! We've now implemented the first step of the authentication flow. In the next section, we will write a page for registration before writing the page for login.

Creating the registration page

If a user needs login credentials, they will need to register first. In this section, we will create a registration form while also handling the necessary requests.

Adding a registration page

Let's start by writing code for the form page. We'll start by writing the registration `form` component:

1. Inside the `src` directory, create a new directory called `components` and then create a new directory called `authentication` inside the newly created directory.

   This directory will contain the registration and login forms.

2. Once that's done, create a file called `RegistrationForm.jsx` inside the `authentication` directory:

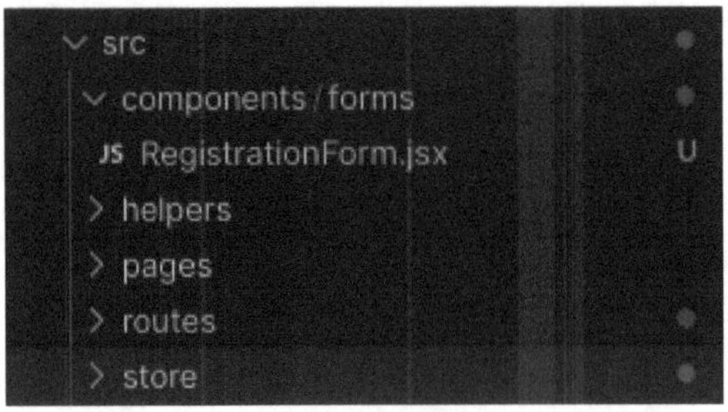

Figure 7.3 – The registration file

React Bootstrap provides `form` components that we can use quickly to create a form and make basic validation. In this component, we'll also have to make a request to the API, register the user details and tokens in the store, and redirect the user to the home page if the request is successful.

3. Next, we will add the needed imports:

src/components/forms/RegistrationForm.js

```
import React, { useState } from "react";
import { Form, Button } from "react-bootstrap";
import axios from "axios";
import { useNavigate } from "react-router-dom";
...
```

4. Now, declare the states and functions we'll use in the component:

src/components/forms/RegistrationForm.js

```
...
function RegistrationForm() {
 const navigate = useNavigate();
 const [validated, setValidated] = useState(false);
 const [form, setForm] = useState({});
 const [error, setError] = useState(null);
...
```

Let's quickly explain what we are doing in the preceding code snippet.

The `navigate` Hook will help us navigate to the home page if the request is successful.

The `validated`, `form`, and `error` states are respectively used to check whether the form is valid or not, the values of each field in the form, and the error message to display if the request doesn't pass.

5. Great! Let's write the function that will handle the form submission:

src/components/forms/RegistrationForm.js

```
...
const handleSubmit = (event) => {
   event.preventDefault();
   const registrationForm = event.currentTarget;

   if (registrationForm.checkValidity() === false) {
     event.stopPropagation();
   }
```

```
    setValidated(true);

    const data = {
      username: form.username,
      password: form.password,
      email: form.email,
      first_name: form.first_name,
      last_name: form.last_name,
      bio: form.bio,
    };
...
```

6. The next step is to use `axios` to make a `POST` request to the API:

src/components/forms/RegistrationForm.js

```
    axios
      .post("http://localhost:8000/api/auth/register/",
             data)
      .then((res) => {
        // Registering the account and tokens in the
        // store

        localStorage.setItem("auth", JSON.stringify({
          access: res.data.access,
          refresh: res.data.refresh,
          user: res.data.user,
        }));

        navigate("/");
      })
      .catch((err) => {
        if (err.message) {
          setError(err.request.response);
        }
      });
  };
```

In the preceding code block, we are first blocking the default form submission behavior with `event.preventDefault()` – that is, reloading the page. Next, we are checking whether the basic validations for the fields are done. With the validation successfully done, we can easily make a request with `axios` and store tokens and user details in `localStorage`.

This way, the user is navigated to the home page.

7. Now, let's add the basic UI components:

src/components/forms/RegistrationForm.js

```
...
return (
    <Form
      id="registration-form"
      className="border p-4 rounded"
      noValidate
      validated={validated}
      onSubmit={handleSubmit}
    >
      <Form.Group className="mb-3">
        <Form.Label>First Name</Form.Label>
        <Form.Control
          value={form.first_name}
          onChange={(e) => setForm({ ...form,
            first_name: e.target.value })}
          required
          type="text"
          placeholder="Enter first name"
        />
        <Form.Control.Feedback type="invalid">
          This file is required.
        </Form.Control.Feedback>
      </Form.Group>
...
```

There is more code after this, but let's grasp the logic here first; the other will be significantly easier.

React Bootstrap provides a `Form` component that we can use to create fields.

`Form.Control` is a component input and it takes as props (`name`, `type`, etc.) attributes any input can take. `Form.Control.Feedback` will show errors when the fields are not valid.

8. Let's do the same for the `last_name` and the `username` fields:

src/components/forms/RegistrationForm.js

```
...
      <Form.Group className="mb-3">
        <Form.Label>Last name</Form.Label>
        <Form.Control
          value={form.last_name}
          onChange={(e) => setForm({ ...form,
            last_name: e.target.value })}
          required
          type="text"
          placeholder="Enter last name"
        />
        <Form.Control.Feedback type="invalid">
          This file is required.
        </Form.Control.Feedback>
      </Form.Group>
      <Form.Group className="mb-3">
        <Form.Label>Username</Form.Label>
        <Form.Control
          value={form.username}
          onChange={(e) => setForm({ ...form, username:
            e.target.value })}
          required
          type="text"
          placeholder="Enter username"
        />
        <Form.Control.Feedback type="invalid">
          This file is required.
        </Form.Control.Feedback>
      </Form.Group>
...
```

9. Let's also add a field for email:

src/components/forms/RegistrationForm.js

```
...
      <Form.Group className="mb-3">
        <Form.Label>Email address</Form.Label>
        <Form.Control
          value={form.email}
          onChange={(e) => setForm({ ...form, email:
            e.target.value })}
          required
          type="email"
          placeholder="Enter email"
        />
        <Form.Control.Feedback type="invalid">
          Please provide a valid email.
        </Form.Control.Feedback>
      </Form.Group>
...
```

10. Let's also add a field for the password:

```
...
      <Form.Group className="mb-3">
        <Form.Label>Password</Form.Label>
        <Form.Control
          value={form.password}
          minLength="8"
          onChange={(e) => setForm({ ...form, password:
            e.target.value })}
          required
          type="password"
          placeholder="Password"
        />
        <Form.Control.Feedback type="invalid">
          Please provide a valid password.
        </Form.Control.Feedback>
      </Form.Group>
...
```

11. Let's add the bio field too. We'll use the `Textarea` field type here:

src/components/forms/RegistrationForm.js

```
...
        <Form.Group className="mb-3">
          <Form.Label>Bio</Form.Label>
          <Form.Control
            value={form.bio}
            onChange={(e) => setForm({ ...form, bio:
              e.target.value })}
            as="textarea"
            rows={3}
            placeholder="A simple bio ... (Optional)"
          />
        </Form.Group>
...
```

12. Finally, add the submit button and export the component:

src/components/forms/RegistrationForm.js

```
...
        <div className="text-content text-danger">
            {error && <p>{error}</p>}
        </div>

        <Button variant="primary" type="submit">
          Submit
        </Button>
      </Form>
    );
  }

  export default RegistrationForm;
```

`RegistrationForm` is now created with the required fields and the logic to handle the form submission.

In the next section, we will add this registration form component to a page and register this page in our application route.

Registering the registration page route

Follow these steps to register the registration page route:

1. Inside the `src/pages` directory, create a file called `Registration.jsx`:

src/pages/Registration.js

```
import React from "react";
import { Link } from "react-router-dom";
import RegistrationForm from "../components/forms/
RegistrationForm";

function Registration() {
  return (
    <div className="container">
      <div className="row">
        <div className="col-md-6 d-flex align-items-
center">
          <div className="content text-center px-4">
            <h1 className="text-primary">
              Welcome to Postman!
            </h1>
            <p className="content">
              This is a new social media site that will
              allow you to share your thoughts and
              experiences with your friends. Register now
              and start enjoying! <br />
              Or if you already have an account, please{"
"}
              <Link to="/login/">login</Link>.
            </p>
          </div>
        </div>
        <div className="col-md-6 p-5">
          <RegistrationForm />
        </div>
      </div>
    </div>
  );
}

export default Registration;
```

We've added simple introduction text to the page and imported the `LoginForm` component.

2. Next, open `App.js` and register the page:

src/App.js

```
...
import Registration from "./pages/Registration";

function App() {
 return (
   <Routes>
     ...
     <Route path="/register/" element={<Registration />}
/>
   </Routes>
 );
}
...
```

3. Great! Now, go to `http://localhost:3000/register/`, and you should have a similar result to this:

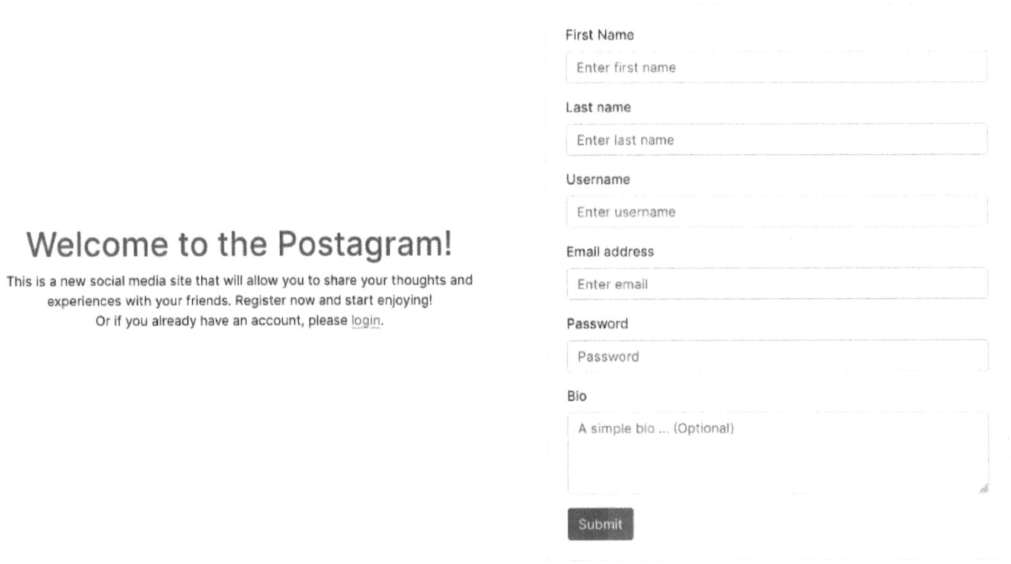

Figure 7.4 – The registration page

4. Test it and register with an account. You'll be redirected to the home page:

Figure 7.5 – The home page

Great! We've just written the registration page.

In the next section, we will create the login page.

Creating the login page

As we have already created the registration page, the logic for login will be pretty similar but with fewer fields.

Adding the login page

Follow these steps to add a login page:

1. Inside the `src/components/authentication` directory, add a new file called `LoginForm.jsx`. This file will contain the form component to log in a user.
2. Next, add the imports:

src/components/authentication/LoginForm.jsx

```
import React, { useState } from "react";
import { Form, Button } from "react-bootstrap";
import axios from "axios";
import { useNavigate } from "react-router-dom";
...
```

3. Write the logic to handle the login:

src/components/authentication/LoginForm.jsx

```
...
function LoginForm() {
  const navigate = useNavigate();
  const [validated, setValidated] = useState(false);
  const [form, setForm] = useState({});
  const [error, setError] = useState(null);

  const handleSubmit = (event) => {
    event.preventDefault();
    const loginForm = event.currentTarget;

    if (loginForm.checkValidity() === false) {
      event.stopPropagation();
    }

    setValidated(true);

    const data = {
      username: form.username,
      password: form.password,
    };
...
```

4. As we did for the registration process, we will now make a request on the login endpoint:

src/components/authentication/LoginForm.jsx

```
...
    axios
      .post("http://localhost:8000/api/auth/login/",
            data)
      .then((res) => {
        // Registering the account and tokens in the
```

```
            // store

            localStorage.setItem("auth", JSON.stringify({
              access: res.data.access,
              refresh: res.data.refresh,
              user: res.data.user,
            }));

            navigate("/");
          })
          .catch((err) => {
            if (err.message) {
              setError(err.request.response);
            }
          });
...
```

This is nearly the same logic as the registration, but here, we are only working with the username and the password.

5. With the logic ready to handle the request made for login, let's add the UI:

src/components/authentication/LoginForm.jsx

```
...
return (
   <Form
     id="registration-form"
     className="border p-4 rounded"
     noValidate
     validated={validated}
     onSubmit={handleSubmit}
   >
     <Form.Group className="mb-3">
       <Form.Label>Username</Form.Label>
       <Form.Control
         value={form.username}
         onChange={(e) => setForm({ ...form, username:
```

```
                    e.target.value })}
      required
      type="text"
      placeholder="Enter username"
    />
    <Form.Control.Feedback type="invalid">
      This file is required.
    </Form.Control.Feedback>
  </Form.Group>
  ...
```

In the preceding code, we are creating the form and adding the first input of the form, the username input.

6. Let's also add the password form input and the submit button:

```
  ...
  <Form.Group className="mb-3">
    <Form.Label>Password</Form.Label>
    <Form.Control
      value={form.password}
      minLength="8"
      onChange={(e) => setForm({ ...form, password:
                    e.target.value })}
      required
      type="password"
      placeholder="Password"
    />
    <Form.Control.Feedback type="invalid">
      Please provide a valid password.
    </Form.Control.Feedback>
  </Form.Group>

  <div className="text-content text-danger">
    {error && <p>{error}</p>}</div>

  <Button variant="primary" type="submit">
    Submit
```

```
        </Button>
      </Form>
  );
}

export default LoginForm;
...
```

We have created the `LoginForm` component with the required fields and logic to handle data submission.

In the next section, we will add `LoginForm` to a page and register this page in the application routes.

Registering the login page

Follow these steps to register the login page:

1. Inside the `src/pages` directory, create a file called `Login.jsx`:

src/pages/Login.jsx

```
import React from "react";
import { Link } from "react-router-dom";
import LoginForm from "../components/forms/LoginForm";
...
```

2. Next, let's add the UI:

src/pages/Login.jsx

```
...
function Login() {
 return (
   <div className="container">
     <div className="row">
       <div className="col-md-6 d-flex
         align-items-center">
         <div className="content text-center px-4">
           <h1 className="text-primary">
             Welcome to Postagram!</h1>
           <p className="content">
```

```
                Login now and start enjoying! <br />
                Or if you don't have an account, please{" "}
                <Link to="/register/">register</Link>.
              </p>
            </div>
          </div>
          <div className="col-md-6 p-5">
            <LoginForm />
          </div>
        </div>
      </div>
    );
  }

  export default Login;
```

This is also quite similar to the registration page.

3. Register the page in the routes of the application in the `App.js` file:

src/App.js

```
...
     <Route path="/login/" element={<Login />} />
...
```

4. Visit `http://localhost:3000/login/`, and you should have a similar page to this:

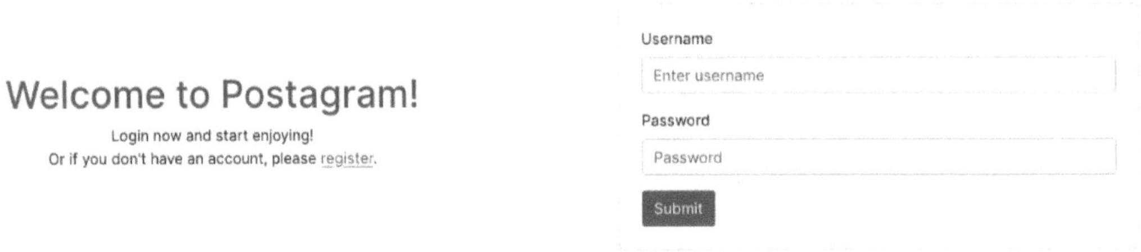

`Figure 7.6 – The login page

5. Test it again, and you should be redirected to the home page.

The authentication flow is working like a charm, but we have some repeated code in our project. Let's do some refactoring by doing a little exercise in the next section.

Refactoring the authentication flow code

Instead of repeating the same code across the code base, we can follow the **Don't Repeat Yourself** (**DRY**) rule. For example, we use the same code to store tokens and user information for the `LoginForm` and `RegistrationForm` components. In this section, we will write a custom React Hook to handle this logic, but before doing that, let's understand what a Hook is.

What is a Hook?

Hooks were first introduced in React 16.8, allowing developers to use more of React's features without writing a class. An interesting example of a React Hook is `useState`.

`useState` is a replacement for `setState`, used inside functional components to manage the internal state of a component. In `LoginForm`, we used `useState` to handle the form values. We also used `useState` to set the message error if the login request returns an error. For a simple test, go to the login page and enter the wrong credentials, and you'll likely get a similar error to this:

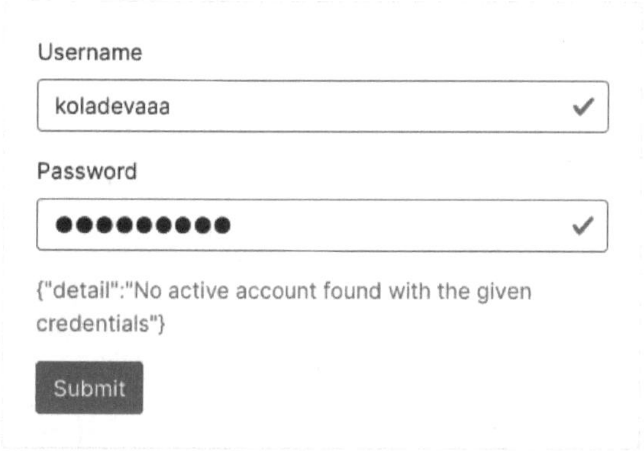

Figure 7.7 – The login form

The logic from this comes from the following lines in `LoginForm.jsx`:

src/authentication/LoginForm.jsx

```
  const [error, setError] = useState(null);
...
      .catch((err) => {
        if (err.message) {
          setError(err.request.response);
        }
      });
```

This is an example of the `useState` Hook, and not every Hook works the same way. For example, you can check the usage of the `useNavigate` Hook in the `LoginForm` component. There are a few rules for using Hooks as per the React documentation:

- **Only call Hooks at the top level**: Don't call Hooks inside loops, conditions, or nested routes
- **Only call Hooks from React functions**: Call Hooks from React function components and custom Hooks

React allows us to write custom Hooks. Let's write a custom Hook to handle user authentication. Inside a new file, we'll write functions that make it easier to retrieve and manipulate the `auth` object in `localStorage`.

Writing code for a custom Hook

Follow these steps to create a custom Hook:

1. Inside the `src` directory, create a new directory called `hooks`. This directory will contain all the Hooks that we'll write in this book.
2. Inside the newly created directory, add a file called `user.actions.js`.
3. Let's add all the necessary content, starting with the imports:

src/hooks/user.actions.js

```
import axios from "axios";
import { useNavigate } from "react-router-dom";
```

4. Next, let's add a function called `useUserActions`. A custom Hook is a JavaScript function whose name starts with `use`:

src/hooks/user.actions.js

```
function useUserActions() {
 const navigate = useNavigate();
 const baseURL = "http://localhost:8000/api";

 return {
   login,
   register,
   logout,
 };
}
```

We can now add the `login` and `logout` functions. These functions will return `Promise`, which, if successful, will register the user data in `localStorage` and redirect the user to the home page, or allow us to catch and handle errors.

5. We will now write the `register` function as a bit of exercise, but it's not that different from the `login` function:

src/hooks/user.actions.js

```
...
  // Login the user
  function login(data) {
    return axios.post(`${baseURL}/auth/login/`,
                      data).then((res) => {
      // Registering the account and tokens in the
      // store
      setUserData(data);
      navigate("/");
    });
  }
...
```

6. Next, write the `logout` function. This function will remove the `auth` item from `localStorage` and redirect the user to the login page:

src/hooks/user.actions.js

```
...
  // Logout the user
  function logout() {
    localStorage.removeItem("auth");
    navigate("/login");
  }
...
```

Note that we are using a method called `setUserData`, which we have not declared yet.

7. After the `useUserActions` function, let's add other utils functions that can be used across the project. These functions will help us to retrieve access tokens, refresh tokens, user information, or set user data:

src/hooks/user.actions.js

```
// Get the user
function getUser() {
  const auth =
    JSON.parse(localStorage.getItem("auth"));
  return auth.user;
}

// Get the access token
function getAccessToken() {
  const auth =
    JSON.parse(localStorage.getItem("auth"));
  return auth.access;
}

// Get the refresh token
function getRefreshToken() {
  const auth =
    JSON.parse(localStorage.getItem("auth"));
  return auth.refresh;
}

// Set the access, token and user property
function setUserData(data) {
  localStorage.setItem(
    "auth",
    JSON.stringify({
      access: res.data.access,
      refresh: res.data.refresh,
      user: res.data.user,
    })
  );
}
```

> **Important note**
>
> You might find it confusing to declare functions after calling them. Writing functions in JavaScript using the `function` keyword allows hoisting, meaning that functions declaration is moved to the top of their scope before code execution. You can learn more at `https://developer.mozilla.org/en-US/docs/Glossary/Hoisting`.

With the functions for retrieving a user, the access and refresh tokens, and the function to set user data in `localStorage`, we can now call the function in the `LoginForm` and `RegisterForm` components.

Using the functions in code

We have a useful Hook, `useUserActions`, in the `user.actions.js` file. We will use this Hook to call the `login` method, thus replacing the old login logic in the `LoginForm.js` file. Let's start by using the newly written custom Hook in the `LoginForm` component. Follow these steps:

1. First, import the Hooks and declare a new variable:

```
...
import { useUserActions } from "../../hooks/user.
actions";

function LoginForm() {
 const [validated, setValidated] = useState(false);
 const [form, setForm] = useState({});
 const [error, setError] = useState(null);
 const userActions = useUserActions();
...
```

2. Now, we can make some changes to the `handleSubmit` function concerning the login request on the API:

src/hooks/user.actions.js

```
const data = {
     username: form.username,
     password: form.password,
   };

   userActions.login(data)
     .catch((err) => {
       if (err.message) {
```

```
            setError(err.request.response);
        }
    });
};
```

In the preceding code block, we did some quick refactoring by removing the old logic for login and setting user data in `localStorage`. The same logic can be applied to `RegistrationForm` (the `register` method is already available in the `useUserActions` Hook). You can modify the `RegistrationForm` component as a small exercise. Feel free to check the code at `https://github.com/PacktPublishing/Full-stack-Django-and-React/blob/chap7/social-media-react/src/components/authentication/RegistrationForm.jsx` to make sure your solution is valid.

3. Great! Let's now use the other utils functions in the `axios` helper and the `ProtectedRoute` component:

src/routes/ProtectedRoute.jsx

```
...
function ProtectedRoute({ children }) {
 const user = getUser();
 return user ? <>{children}</> : <Navigate
   to="/login/" />;
...
```

4. Next, let's do some tweaks in the `axios` helper:

```
...
import { getAccessToken, getRefreshToken } from "../
hooks/user.actions";
...
 config.headers.Authorization = `Bearer
${getAccessToken()}`;
...
   .post("/refresh/token/", null, {
     baseURL: "http://localhost:8000",
     headers: {
       Authorization: `Bearer ${getRefreshToken()}`,
...
     const { access, refresh, user } = resp.data;
     failedRequest.response.config.headers[
       "Authorization"] =
```

```
        "Bearer " + access;
      localStorage.setItem("auth", JSON.stringify({
        access, refresh, user }));
    })
    .catch(() => {
      localStorage.removeItem("auth");
    });
...
```

In the preceding code block, we used the `getAccessToken` and `getRefreshToken` functions to retrieve the access token and the refresh token from `localStorage` for the requests. We just replaced the old logic to retrieve the access and refresh tokens.

And we are done. We have a pure React logic for the authentication flow, which will help us manage the CRUD operations for the posts and comments in the following chapters.

Summary

In this chapter, we dived deeper into more concepts, such as authentication in a React application. We implemented a clean logic for requests on the Django API with access tokens and also implemented the refresh logic if the access token has expired. We also had the chance to use more of the Bootstrap components to not only style login and registration forms but also to create login and register pages. Finally, we implemented a custom React Hook to handle everything concerning authentication on the frontend, with methods for registration and login, and some utilities to retrieve tokens from `localStorage` and also set tokens and user data in `localStorage`. The creation of the custom Hook helped us make some refactoring in the code base according to the DRY principle.

In the next chapter, we will allow users to create posts from the React application. We will learn how to make requests to the backend using the custom-written `axiosService`, display modals, handle more complex React states, and also use the `useContext` React Hook to handle pop-up displays.

Questions

1. What is `localStorage`?
2. What is React-Router?
3. How do you configure a protected route in React?
4. What is a React Hook?
5. Give three examples of React Hooks.
6. What are the two rules of React Hooks?

8
Social Media Posts

Social media already has authentication added on the frontend side. We can now authenticate the user through registration or login, fetch the user data, and show it. Now that we can store JWT tokens, we can make requests to the API for any protected resources, and we will start with the `post` resource.

In this chapter, we'll focus on **CRUD** operations on posts. We'll implement listing, creating, updating, and deleting post features. You will learn how to create and manage a Modal in React, how to handle a form from validation to submission, and how to design and integrate components into a React page.

This chapter will cover the following topics:

- Listing posts in a feed
- Creating a post using a form
- Editing and deleting a post
- Liking a post

Technical requirements

Make sure to have VS Code and an updated browser installed and configured on your machine. You can find the code of this chapter at `https://github.com/PacktPublishing/Full-stack-Django-and-React/tree/chap8`.

Creating the UI

The REST API is ready to accept requests and list the API. For the next steps, ensure that the Django server is running on the machine at port `localhost:8000`. The first step is implementing a post feed with a ready design and UI. Before coding the components for reading, creating, updating, and deleting a component, we need to analyze the UI and also make sure we have the right configurations and components to ease the development with React. We will mostly build the navigation bar and the layout.

Here's the feed UI of the home page:

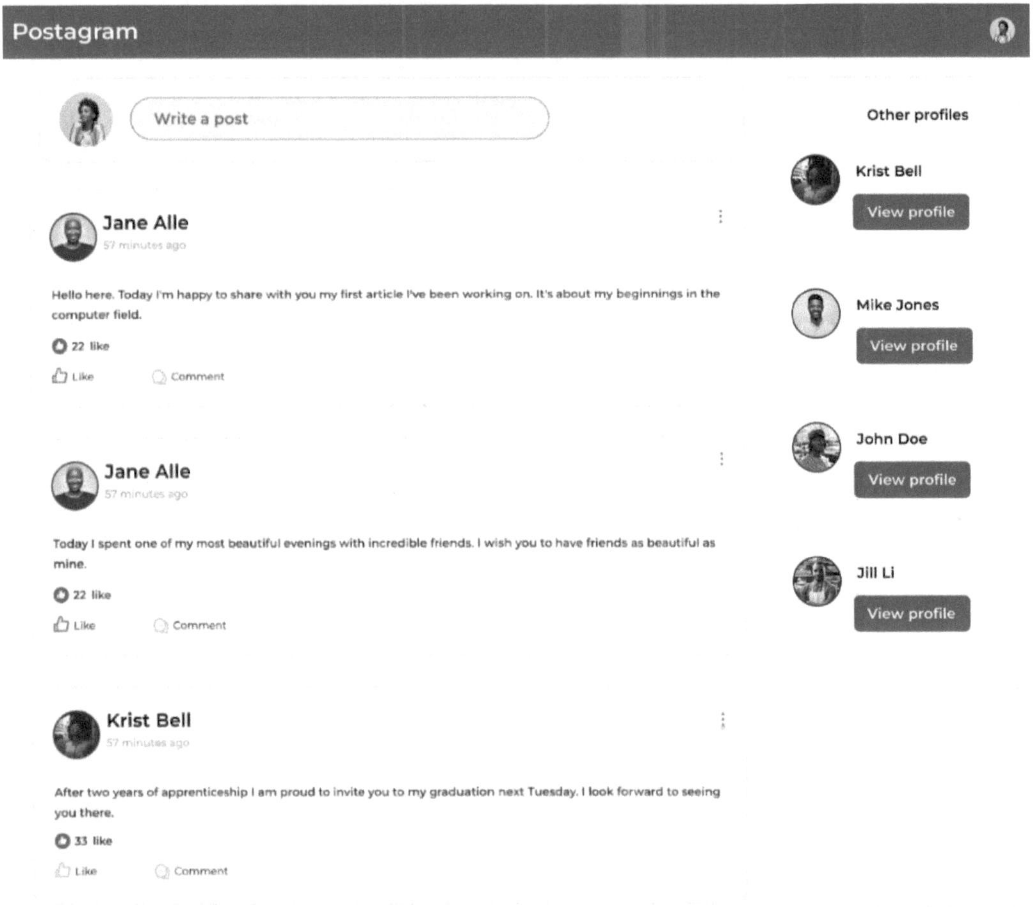

Figure 8.1 – Feed UI wireframe

In the following figure, we have another illustration representing the UI and the page's structure. We are using flex columns, and we'll use Bootstrap flex components to design the page:

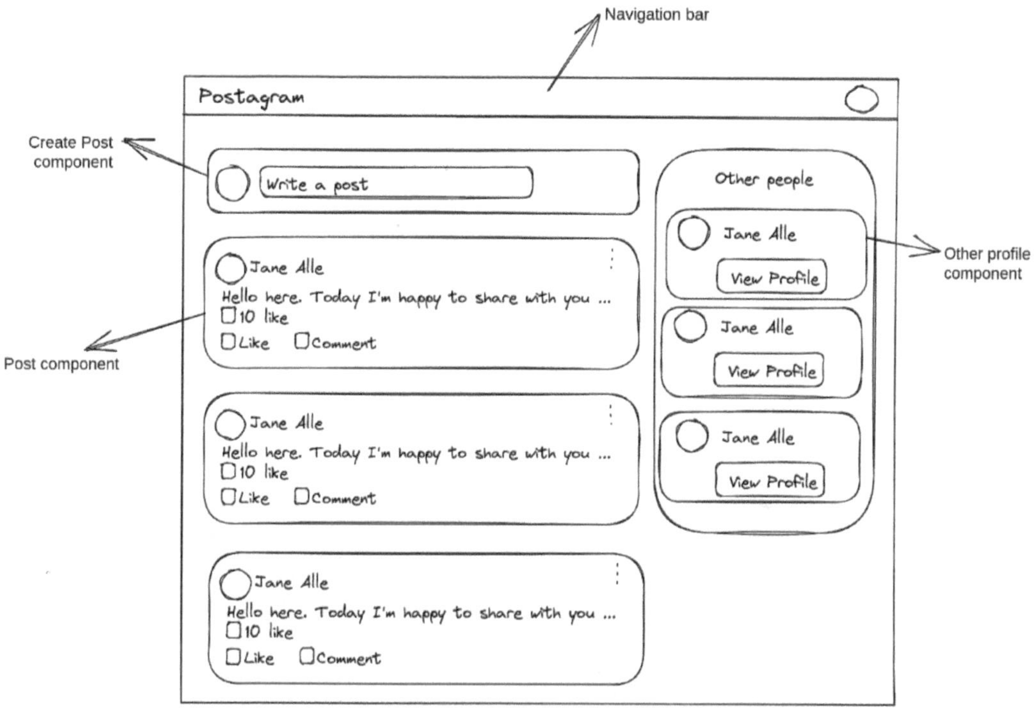

Figure 8.2 – Wireframe

The navigation bar will be available on other pages of the React application, and by making the navigation bar a component, it'll be reused. We can make the integration of the navigation bar easier by having a `Layout` component that will be used when building the pages. Let's start by adding the navigation bar component.

Adding the NavBar component

The `NavBar` component, or the navigation bar component, should help to quickly navigate the UI. Here's an screenshot of the `NavBar` component:

Figure 8.3 – Navbar

The NavBar will have three links:

- A link to redirect to the feed page (**1**)
- A link to redirect to the profile page (**2**)
- A link to log out (**3**)

Here's a simple wireframe to better illustrate where the links will go.

Figure 8.4 – Wireframe of the navbar

Let's add the component. Follow these steps to do so:

1. Inside the `src/components/` directory, add a new file called `Navbar.jsx`. This file will contain the code for the `NavBar` component. Bootstrap already provides a `NavBar` component we can use. Let's start with the component definition and the necessary imports:

src/components/Navbar.jsx

```
import React from "react";
import { randomAvatar } from "../utils";
import { Navbar, Container, Image, NavDropdown, Nav }
from "react-bootstrap";
import { useNavigate } from "react-router-dom";
...
```

2. With the already written function, we can add the `NavBar` component and style it. `react-bootstrap` provides components that we can use to make the coding of our components faster. The props that the components require make the customization of these components easier:

src/components/Navbar.jsx

```
...
function Navigationbar() {
 return (
   <Navbar bg="primary" variant="dark">
     <Container>
       <Navbar.Brand className="fw-bold" href="#home">
         Postagram
```

```
          </Navbar.Brand>
          <Navbar.Collapse
            className="justify-content-end">
            <Nav>
              <NavDropdown
                title={
                  <Image
                    src={randomAvatar()}
                    roundedCircle
                    width={36}
                    height={36}
                  />
                }
              >
                <NavDropdown.Item href="#">Profile
                </NavDropdown.Item>
                <NavDropdown.Item onClick={handleLogout}>
                  Logout</NavDropdown.Item>
              </NavDropdown>
            </Nav>
          </Navbar.Collapse>
        </Container>
      </Navbar>
  );
}

export default Navigationbar;
```

3. Let's add the function that handles the logout:

src/components/Navbar.jsx

```
...
function Navigationbar() {
  const navigate = useNavigate();
  const handleLogout = () => {
    localStorage.removeItem("auth");
```

```
        navigate("/login/");
    };
...
```

I will use a website that generates random avatars for the avatar. In the next chapter, we'll do a little exercise to add an upload profile picture feature, but the image generator will do the work for the moment.

4. In the `src` directory, add a new file called `utils.js`. This file will contain functions that we'll reuse in the React application:

src/utils.js

```
export const randomAvatar = () =>
 `https://i.pravatar.cc/300?img=${Math.floor(Math.
random() * 60) + 1}`;
```

The `pravatar` service supports parameters in the URL and has over 60 images. We are using the Math library to generate a random number representing the image's ID. We can now write the `Layout` component.

Adding the Layout component

A good React project has visual consistency but should also come with less repetition of code. For example, the navigation bar on this React project will be present on the home page but also on the profile page. When developing in HTML and CSS directly, we would have repeated the same piece of code for the navigation bar, but we can avoid the repetition using React by creating a `Layout` component.

In the `src/components` directory, add a file called `Layout.jsx`. This file will contain the code for the `Layout` component:

src/components/Layout.jsx

```
import React from "react";
import Navigationbar from "./Navbar";

function Layout(props) {
 return (
   <div>
     <Navigationbar />
```

```
      <div className="container m-5">{props.children}</div>
    </div>
  );
}

export default Layout;
```

We have a new syntax here: `children`. In React, `children` is used for displaying whatever you include between the opening and closing tags when invoking a component. Here's a simple example with an image component:

```
const Picture = (props) => {
  return (
    <div>
      <img src=""/>
      {props.children}
    </div>
  )
}
```

The component can then be used, and we can add content or other components:

```
render () {
  return (
    <div className='container'>
      <Picture>
          <p>This a children element.</p>
      </Picture>
    </div>
  )
}
```

Whenever the `Picture` component is invoked, `props.children` will also be displayed, which is just a reference to the component's opening and closing tags. In our context, `props.children` will contain mostly the content of the pages of the React application.

For example, on the home page, we have posts and profiles listed; these elements will be children of the `Layout` component. Without further ado, let's use the `Layout` component.

Using the Layout component on the home page

Inside `Home.jsx`, we'll rewrite the code to use the `Layout` component. Here's the new code:

src/pages/Home.jsx

```
import React from "react";
import Layout from "../components/Layout";

function Home() {

 return (
   <Layout>

   </Layout>
 );
}

export default Home;
```

Great. Let's start by adding the input to create a new post, as shown in *Figure 8.2*.

Creating a post

To create and add posts, follow these steps:

1. In `src/components`, add a new directory called `posts`. This directory will contain all components used for the post feature. We'll have components to create a post, display a post, and update a post.
2. Inside the newly created directory, add a file called `CreatePost.jsx`. This file will contain the code for the logic and the UI to make a post.
3. What we have here is a UI component called `Modal`. `react-bootstrap` provides a modal-ready element that we can easily customize for our needs. Let's start by adding the needed imports and defining the component function:

   src/components/post/CreatePost.jsx

   ```
   import React, { useState } from "react";
   import { Button, Modal, Form } from "react-bootstrap";
   import axiosService from "../../helpers/axios";
   ```

```
import { getUser } from "../../hooks/user.actions";

function CreatePost()
  return ()
};
export default CreatePost;
```

4. The input for the post creation will be within the `Modal` component. As we did earlier, we will also add methods and state management for the form. But first, let's write the modal and the clickable input:

src/components/post/CreatePost.jsx

```
...
function CreatePost() {
 const [show, setShow] = useState(false);
 const handleClose = () => setShow(false);
 const handleShow = () => setShow(true);

 return (
   <>
     <Form.Group className="my-3 w-75">
       <Form.Control
         className="py-2 rounded-pill border-primary
                    text-primary"
         type="text"
         placeholder="Write a post"
         onClick={handleShow}
       />
     </Form.Group>

     {/*Add modal code here*/}
   </>
 );
}

export default CreatePost;
```

5. We are first adding the input that will trigger the Modal to be displayed. A click on the modal will set the `show` state to `True`, the state that is used for opening the modal. Let's add the code for the modal:

src/components/post/CreatePost.jsx

```
...
      <Modal show={show} onHide={handleClose}>
        <Modal.Header closeButton className="border-0">
          <Modal.Title>Create Post</Modal.Title>
        </Modal.Header>
        <Modal.Body className="border-0">
          <Form noValidate validated={validated}
            onSubmit={handleSubmit}>
            <Form.Group className="mb-3">
              <Form.Control
                name="body"
                value={form.body}
                onChange={(e) => setForm({ ...form,
                  body: e.target.value })}
                as="textarea"
                rows={3}
              />
            </Form.Group>
          </Form>
        </Modal.Body>
        <Modal.Footer>
          <Button variant="primary"
            onClick={handleSubmit}
            disabled={form.body === undefined}>
            Post
          </Button>
        </Modal.Footer>
      </Modal>
...
```

6. The UI for the modal is created. We need now to add the `handleSubmit` function and the other logic for the form handling:

src/components/post/CreatePost.jsx

```
function CreatePost() {
...
  const [validated, setValidated] = useState(false);
  const [form, setForm] = useState({});

  const user = getUser();

...

  const handleSubmit = (event) => {
    event.preventDefault();
    const createPostForm = event.currentTarget;

    if (createPostForm.checkValidity() === false) {
      event.stopPropagation();
    }

    setValidated(true);

    const data = {
      author: user.id,
      body: form.body,
    };

    axiosService
      .post("/post/", data)
      .then(() => {
        handleClose();
        setForm({});
      })
      .catch((error) => {
        console.log(error);
      });
```

```
};
...
```

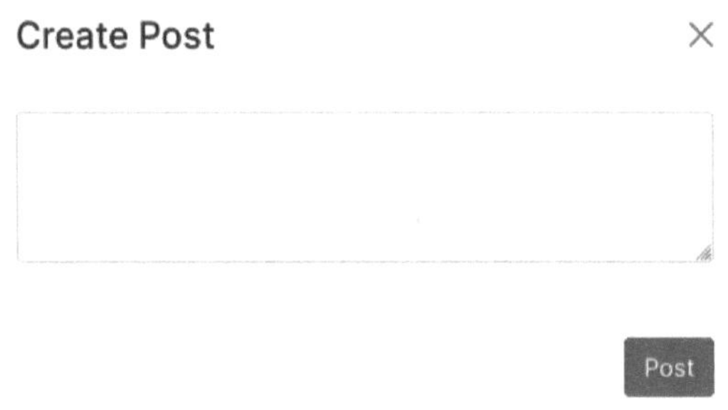

Figure 8.5 – Create Post component

We are nearly done, but we need an essential feature for every action, such as form handling. We need to send feedback to the user to tell them whether their request has passed. In our context, when a user creates a post, we'll show a success toast or an error toast:

Figure 8.6 – A successful toast

The toast will be reused for post deletion and also updating. It will also be used for comment creation, modification, and deletion, as well as the profile modification that we will add later. We will add the `Toast` component in the next section.

Adding the Toast component

Let's quickly create a component called `Toaster` that we will use show toast in the React application.

In `src/components`, create a new file called `Toaster.jsx`. This file will contain the code for the `Toaster` component:

src/components/Toaster.jsx

```
import React from "react";
import { Toast, ToastContainer } from "react-bootstrap";
```

```
function Toaster(props) {
  const { showToast, title, message, onClose, type } =
    props;

  return (
    <ToastContainer position="top-center">
      <Toast onClose={onClose} show={showToast} delay={3000}
        autohide bg={type}>
        <Toast.Header>
          <strong className="me-auto">{title}</strong>
        </Toast.Header>
        <Toast.Body>
          <p className="text-white">{message}</p>
        </Toast.Body>
      </Toast>
    </ToastContainer>
  );
}

export default Toaster;
```

The `Toaster` component takes some props:

- `showToast`: The Boolean that is used to show the toast or not. Ideally, depending on the output we receive from a request on the server, we'll set the state to `true`, which will show the toast.
- `title`: This represents the title of the toast.
- `message`: This conveys the message we'll be showing in the toast.
- `onClose`: The function that handles the closing of the toast. This function is essential; otherwise, the toast will never disappear.
- `type`: This represents the type of toast to show. In our context, we'll either use `success` or `danger`.

Let's import this component in `CreatePost.jsx` and use it.

Adding toaster to post creation

In the `CreatePost.jsx` file, we will add new states that we will pass as props to the `Toaster` component:

src/components/post/CreatePost.jsx

```
...
import Toaster from "../Toaster";

function CreatePost() {
...
  const [showToast, setShowToast] = useState(false);
  const [toastMessage, setToastMessage] = useState("");
  const [toastType, setToastType] = useState("");
...
  const handleSubmit = (event) => {
    ...

    axiosService
      .post("/post/", data)
      .then(() => {
        handleClose();
        setToastMessage("Post created 🚀");
        setToastType("success");
        setForm({});
        setShowToast(true);
      })
      .catch(() => {
        setToastMessage("An error occurred.");
        setToastType("danger");
      });
  };
```

We can import the `Toaster` component and pass the newly added states as props:

src/components/post/CreatePost.jsx

```
...
    </Modal>
```

```
        <Toaster
          title="Post!"
          message={toastMessage}
          showToast={showToast}
          type={toastType}
          onClose={() => setShowToast(false)}
        />
      </>
...
```

And we are done writing the `CreatePost` component. For the next step, we need to integrate it into the home page.

Adding the CreatePost component to the home page

The `CreatePost` component is ready now, and we can use it. First, import it into the `Home.jsx` file and modify the UI.

The home page will have two parts:

- The first part will contain the list of posts (**1** in *Figure 8.7*)
- The second part will include a list of five profiles (**2** in *Figure 8.7*)

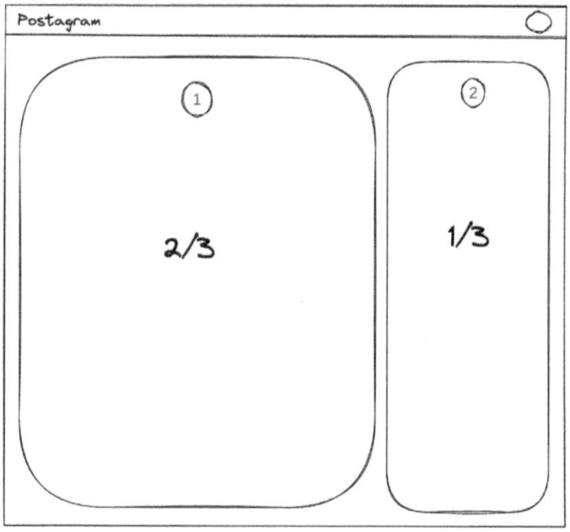

Figure 8.7 – Structure of the home page

We can achieve the result quickly by using rows and columns components provided by `react-bootstrap`. We won't focus on the second part (listing the profiles) for the moment. Let's ensure we have all **CRUD** operations for the post feature:

1. Inside the `Home.jsx` file, add the following content. We'll start by importing and adding the rows:

src/pages/Home.jsx

```
import React from "react";
import Layout from "../components/Layout";
import { Row, Col, Image } from "react-bootstrap";
import { randomAvatar } from "../utils";
import useSWR from "swr";
import { fetcher } from "../helpers/axios";
import { getUser } from "../hooks/user.actions";
import CreatePost from "../components/posts/CreatePost";

function Home() {
 const user = getUser();

 if (!user) {
   return <div>Loading!</div>;
 }

 return (
   <Layout>
     <Row className="justify-content-evenly">
       <Col sm={7}>
         <Row className="border rounded
           align-items-center">
           <Col className="flex-shrink-1">
             <Image
               src={randomAvatar()}
               roundedCircle
               width={52}
               height={52}
               className="my-2"
             />
```

```
                </Col>
                <Col sm={10} className="flex-grow-1">
                  <CreatePost />
                </Col>
              </Row>
            </Col>
          </Row>
        </Layout>
      );
    }

    export default Home;
```

2. Great! Make sure to save the changes, start the server, and go to the home page. You'll have something similar to this:

Figure 8.8 – Create Post UI

3. Click on the input, and a modal will show up. Type anything you want in the input and submit it. The modal will close, and you'll have a toast appearing at the top center of the page:

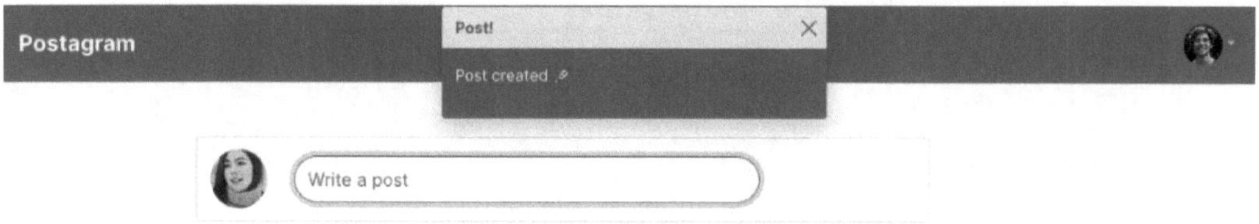

Figure 8.9 – Toast after successful post creation

Great! We can now create posts using our React application. To make it possible, we have created a `Modal` component and a form with React Bootstrap to handle data validation and submission. And because feedback is an important aspect of user experience, we have added a toaster with React Bootstrap and integrated it with the `useContext` Hook to notify the user of the result of the requests.

The next step is to list all the posts and add actions such as deletion and modification.

Listing posts on the home page

Now that users can create posts, we need to list the posts on the home page but also allow the user to access them. This will require the creation of a component to display information about a post. As shown in *Figure 8.1*, under the **Write a post** input, we have a list of posts. The home page structure is already added, so we need to add a component that will handle the logic behind showing information about a post.

Here's the flow to list the posts on the home page:

- We use the `swr` library to fetch a list of posts
- We loop through the list of posts and then pass a post as props to a component called `Post`, which will show data about a post

Before starting to fetch data, let's create the `Post` component.

Writing the Post component

To create a `Post` component, follow these steps:

1. Inside the `src/components/post/` directory, create a new file called `Post.jsx`. This file will contain the logic to show post data and logic such as like or remove like, deletion, and modification. Here's a wireframe of the `Post` component:

Figure 8.10 – Post component

2. To make things faster, we will work with the `Card` component provided by `react-bootstrap`. The `Card` component comes with a structure containing a title, body, and footer:

src/components/post/Post.jsx

```
import React, { useState } from "react";
import { format } from "timeago.js";
import {
  LikeFilled,
  CommentOutlined,
```

```
  LikeOutlined,
} from "@ant-design/icons";
import { Image, Card, Dropdown } from "react-bootstrap";
import { randomAvatar } from "../../utils";

function Post(props) {
  const { post, refresh } = props;

  const handleLikeClick = (action) => {
    axiosService
      .post(`/post/${post.id}/${action}/`)
      .then(() => {
        refresh();
      })
      .catch((err) => console.error(err));
  };

  return (
    <>
      <Card className="rounded-3 my-4">
        {/* Add card body here*/}
      </Card>
    </>
  );
}

export default Post;
```

The `Post` component accepts two props:

- The `post` object containing data about a post.
- The `refresh` function. This function will come from the SWR `posts` object, and SWR returns an object with a `mutate` method that can be used to trigger the fetching of data.

3. We also profited from adding the `handleLikeClick` function. Two actions can be passed to the function: either `like` or `remove_like`. If the request succeeds, we can refresh the posts. Great! Let's start by adding the `Card` body. It'll contain the avatar of the author of the post, the name, and the time elapsed since the creation of the post:

src/components/post/Post.jsx

```
...
<Card.Body>
         <Card.Title className="d-flex flex-row
           justify-content-between">
           <div className="d-flex flex-row">
             <Image
               src={randomAvatar()}
               roundedCircle
               width={48}
               height={48}
               className="me-2 border border-primary
                          border-2"
             />
             <div className="d-flex flex-column
               justify-content-start
               align-self-center mt-2">
               <p className="fs-6 m-0">
                 {post.author.name}</p>
               <p className="fs-6 fw-lighter">
                 <small>{format(post.created)}</small>
               </p>
             </div>
           </div>
         </Card.Title>
       </Card.Body>
...
```

4. Go ahead and add the body of the post and the likes count:

src/components/post/Post.jsx

```
...
          </Card.Title>
          <Card.Text>{post.body}</Card.Text>
          <div className="d-flex flex-row">
            <LikeFilled
              style={{
                color: "#fff",
                backgroundColor: "#0D6EFD",
                borderRadius: "50%",
                width: "18px",
                height: "18px",
                fontSize: "75%",
                padding: "2px",
                margin: "3px",
              }}
            />
            <p className="ms-1 fs-6">
              <small>{post.likes_count} like</small>
            </p>
          </div>
        </Card.Body>
...
```

5. We can now move to the `Card` footer containing the like and comment UI. Let's start by adding the **Like** icon followed by text:

src/components/post/Post.jsx

```
...
        </Card.Body>
        <Card.Footer className="d-flex bg-white w-50
          justify-content-between border-0">
          <div className="d-flex flex-row">
            <LikeOutlined
```

```
            style={{
              width: "24px",
              height: "24px",
              padding: "2px",
              fontSize: "20px",
              color: post.liked ? "#0D6EFD" :
                "#C4C4C4",
            }}
            onClick={() => {
              if (post.liked) {
                handleLikeClick("remove_like");
              } else {
                handleLikeClick("like");
              }
            }}
          />
          <p className="ms-1">
            <small>Like</small>
          </p>
        </div>
  {/* Add comment icon here*/}
      </Card.Footer>
    </Card>
...
```

6. Now go ahead and add the **Comment** icon followed by the text:

src/components/post/Post.jsx

```
...
        <div className="d-flex flex-row">
          <CommentOutlined
            style={{
              width: "24px",
              height: "24px",
              padding: "2px",
```

```
                fontSize: "20px",
                color: "#C4C4C4",
              }}
            />
            <p className="ms-1 mb-0">
              <small>Comment</small>
            </p>
          </div>
        </Card.Footer>
...
```

The Post component is entirely written; we can use it on the home page now.

Adding the Post component to the home page

Let's now add our Post component to the home page.

In the Home.jsx file, import the Post component:

src/pages/Home.jsx

```
...
import { Post } from "../components/posts";
...
```

We can now use the components in the code by first fetching posts from the server:

src/pages/Home.jsx

```
...
function Home() {
  const posts = useSWR("/post/", fetcher, {
    refreshInterval: 10000,
  });
...
```

The useSWR Hook can accept some parameters, such as refreshInterval. Here, the returned data is refreshed every 10 seconds. We can now use these objects in the UI:

src/pages/Home.jsx

```
...
          <Col sm={10} className="flex-grow-1">
            <CreatePost />
          </Col>
        </Row>
        <Row className="my-4">
          {posts.data?.results.map((post, index) => (
            <Post key={index} post={post}
              refresh={posts.mutate} />
          ))}
        </Row>
      </Col>
...
```

Great! After adding the Post component to the home page, you should have a similar result to this:

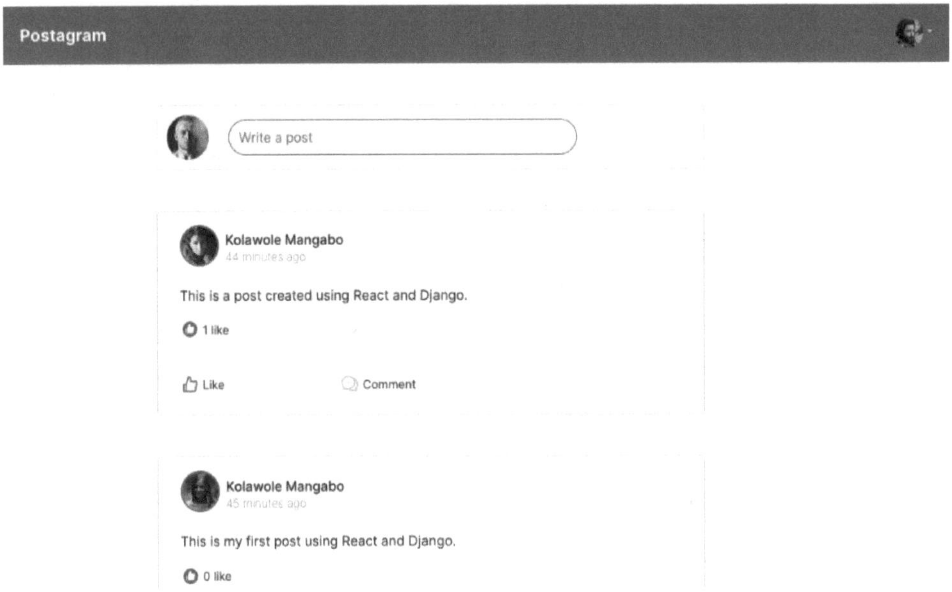

Figure 8.11 – List of posts

You can click on the **Like** icon and see what happens. Usually, the button color will change to blue, and the like count will increase. The behavior will be reversed if you click again on the **Like** icon. But the `Post` component has a **More** dropdown in the top-right corner:

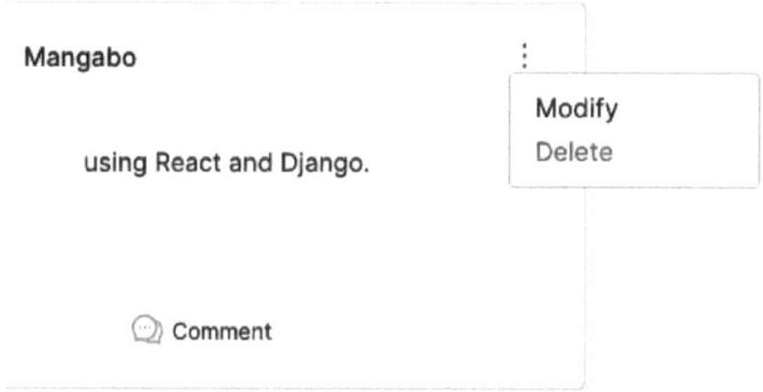

Figure 8.12 – Adding the more dropdown

`react-bootstrap` provides a `Dropdown` component we can use to achieve the same result. In the `Post.jsx` file, import the `Dropdown` component from `react-bootstrap`. As we will add the logic for post deletion, let's also import the `Toaster` component:

src/components/post/Post.jsx

```
import { Button, Modal, Form, Dropdown } from "react-
bootstrap";
import Toaster from "../Toaster";
...
```

We then have to write the component we will pass to the `Dropdown` component as the title:

src/components/post/Post.jsx

```
...
const MoreToggleIcon = React.forwardRef(({ onClick }, ref) => (
  <Link
    to="#"
    ref={ref}
    onClick={(e) => {
      e.preventDefault();
      onClick(e);
    }}
  >
```

```
    <MoreOutlined />
  </Link>
));

function Post(props) {
...
```

We can now add the `Dropdown` component to the UI. We need to make it conditional so that only the author of the post can access these options. We will just retrieve the user from `localStorage` and compare `user.id` to `author.id`:

src/components/post/Post.jsx

```
...
function Post(props) {
...
  const [showToast, setShowToast] = useState(false);
  const user = getUser();
...
  const handleDelete = () => {
    axiosService
      .delete(`/post/${post.id}/`)
      .then(() => {
        setShowToast(true);
        refresh();
      })
      .catch((err) => console.error(err));
  };
  return (
  ...
```

Let's add the component UI and the `Toaster` component:

src/components/post/Post.jsx

```
...
      <Card className="rounded-3 my-4">
        <Card.Body>
          <Card.Title className="d-flex flex-row
```

```
            justify-content-between">
            ...
            {user.name === post.author.name && (
              <div>
                <Dropdown>
                  <Dropdown.Toggle as={MoreToggleIcon}>
                  </Dropdown.Toggle>
                  <Dropdown.Menu>
                    <Dropdown.Item>Update</>
                    <Dropdown.Item
                      onClick={handleDelete}
                      className="text-danger"
                    >
                      Delete
                    </Dropdown.Item>
                  </Dropdown.Menu>
                </Dropdown>
              </div>
            )}
          </Card.Title>
          ...
      </Card>
      <Toaster
        title="Post!"
        message="Post deleted"
        type="danger"
        showToast={showToast}
        onClose={() => setShowToast(false)}
      />
    </>
  );
}

export default Post;
```

The `Dropdown` component is also added to the toaster. Each time a post is deleted, a red toaster will pop up at the top center of the page:

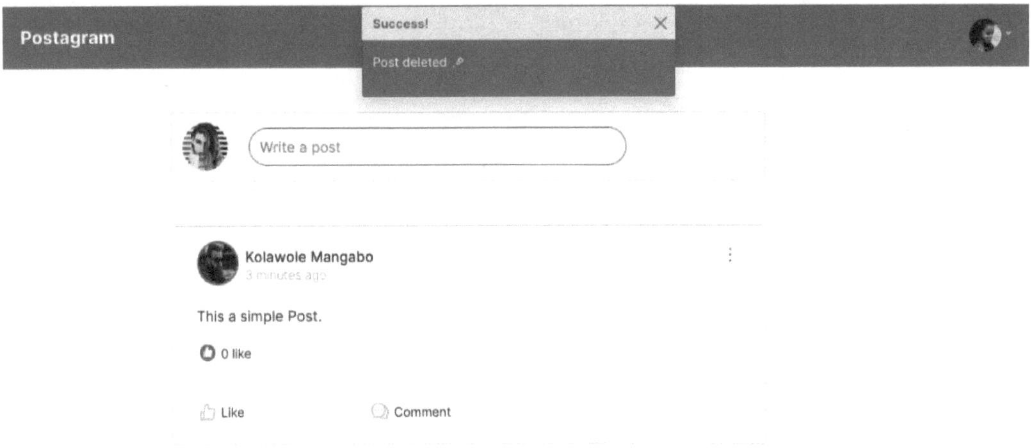

Figure 8.13 – Deleting a post

The user can now delete their own post and the functionality is accessible directly from the `Post` component. We have explored how to use the `UseContex` Hook again and also how to create a dropdown using `react-bootstrap`.

The **CRUD** operations on the post feature are nearly done and only the update feature remains. It's easy, and you will implement it as a small exercise, but I'll add the necessary code and instructions to follow.

Updating a post

As mentioned earlier, the implementation of this feature is a simple exercise. Here's the flow the user will typically follow when modifying a post:

1. Click on the **More** dropdown menu.
2. Select the **Modify** option.
3. A modal is shown with the body of the post, and the user can modify it.
4. Once it's done, the user saves, and the modal is closed.
5. A toast will pop up with the content **Post updated** 🚀.

The feature is similar to `CreatePost.jsx`; the difference is that the `UpdatePost` component will receive a `post` object as props. Here's the skeleton of the code:

src/components/post/UpdatePost.jsx

```
import React, { useState } from "react";
import { Button, Modal, Form, Dropdown } from "react-
bootstrap";
import axiosService from "../../helpers/axios";
import Toaster from "../Toaster";

function UpdatePost(props) {
 const { post, refresh } = props;
 const [show, setShow] = useState(false);

 const handleClose = () => setShow(false);
 const handleShow = () => setShow(true);

 // Add form handling logic here
 return (
   <>
     <Dropdown.Item onClick={handleShow}>Modify
     </Dropdown.Item>

     <Modal show={show} onHide={handleClose}>
     {/*Add UI code here*/}
     </Modal>
   </>
 );
}

export default UpdatePost;
```

The component is called in the `Post.jsx` file and used like this:

src/components/post/Post.jsx

```
...
import UpdatePost from "./UpdatePost";
```

```
...
            </div>
            {user.name === post.author.name && (
              <div>
                <Dropdown>
                  <Dropdown.Toggle as={MoreToggleIcon}>
                  </Dropdown.Toggle>
                  <Dropdown.Menu>
                    <UpdatePost post={post}
                      refresh={refresh} />
                    <Dropdown.Item
                      onClick={handleDelete}
                      className="text-danger"
                    >
                      Delete
                    </Dropdown.Item>
                  </Dropdown.Menu>
                </Dropdown>
              </div>
            )}
...
```

Good luck with the exercise. You can find the solution at `https://github.com/PacktPublishing/Full-stack-Django-and-React/blob/main/social-media-react/src/components/posts/UpdatePost.jsx`.

Minor refactoring

Firstly, there is no refresh made when a new post is created. As we did for the `UpdatePost.jsx` component, we can also pass some props to the `CreatePost` component:

src/pages/Home.jsx

```
...
            <Col sm={10} className="flex-grow-1">
              <CreatePost refresh={posts.mutate} />
            </Col>
...
```

And, we can call the `refresh` method when a post is successfully created:.

src/components/posts/CreatePost.jsx

```
function CreatePost(props) {
  const { refresh } = props;
  ...
    axiosService
      .post("/post/", data)
      .then(() => {
      ...
        setForm({});
        setShowToast(true);
        refresh();
      })
  ...
```

Now, every time a user adds a post, he will see the newly created post on the Home page without the need of reloading the page.

Secondly, the `Toaster` component is created but we need to think about how to call the component in the project. Let's not forget that this component is created to return feedback to the user about a successful or failed request, thus the component should be reusable in the whole project, which is what we've actually done, right?

Well, no, and this is not desirable as it will violate the **DRY** rule. The logic to call the component is repeated across all pages that call this component. How can we resolve this? What if we can have the `Toaster` component higher in the component hierarchy and then be able to call or show the toaster from any child component?

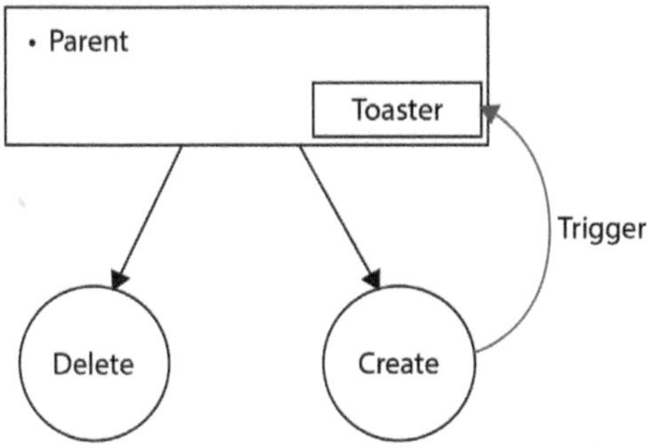

Figure 8.14 – Parent and child components

In the preceding figure, we will be able to trigger the display of a toaster in the project directly in a parent component from a child component (`CreatePost`). React provides an interesting way to manage state across parents and child components and this is called context. **React Context** allows you to share state or modify state across parent and child components more easily. In the `Layout.jsx` file, create a new context using the `createContext` method:

src/components/Layout.jsx

```
import React, { createContext, useMemo, useState } from
"react";

export const Context = createContext("unknown");

function Layout(props) {
```

Then in the `Layout` component scope, let's define the state containing the data that the toaster will use to display information. We will also wrap the component JSX content inside the `Context` component and add a method to modify the state from any child components of the `Layout` component:

src/components/Layout.jsx

```
function Layout(props) {
  ...
  const [toaster, setToaster] = useState({
    title: "",
    show: false,
    message: "",
    type: "",
  });

  const value = useMemo(() => ({ toaster, setToaster }),
                               [toaster]);
  ...
  return (
    <Context.Provider value={value}>
      <div>
        <NavigationBar />
        {hasNavigationBack && (
```

```
        <ArrowLeftOutlined
          style={{
            color: "#0D6EFD",
            fontSize: "24px",
            marginLeft: "5%",
            marginTop: "1%",
          }}
          onClick={() => navigate(-1)}
        />
      )}
      <div className="container my-2">
        {props.children}</div>
    </div>
    <Toaster
      title={toaster.title}
      message={toaster.message}
      type={toaster.type}
      showToast={toaster.show}
      onClose={() => setToaster({ ...toaster, show: false
        })}
    />
  </Context.Provider>
 );
}

export default Layout;
```

In the preceding code, we have introduced a new function Hook called `useMemo`, which helps to memorize the context value (caching the value of the context) and avoid the creation of new objects every time there is a re-rendering of the `Layout` component.

We will then be able to access the `toaster` state and call the `setToaster` function from any child component:

```
const { toaster, setToaster } = useContext(Context);
```

Summary

In this chapter, we've gone deeper into React programming by creating the components needed for the CRUD operations used in the post feature. We have covered concepts such as props passing, parent-children component creation, UI component customization, and modal creation. That led to the partial completion of the home page of the `Postagram` project. We also learned more about the `useState` and `useContext` Hooks and how they affect state in React. We have also learned how to create a `Dropdown` component, how to create a custom toaster, and the importance of layout in a React project.

In the next chapter, we'll focus on the CRUD operations of the comment feature. This will lead us to add a **Profile** page and a **Post** page to display comments. We'll also make simple and quick assessments to add Like features to the comments too.

Questions

1. What is a modal?
2. What is a prop?
3. What is a children element in React?
4. What is a wireframe?
5. What is the map method used in JSX?
6. What is the usage of the `mutate` method on SWR objects?

9
Post Comments

An exciting part of every social media platform is the comment functionality. In the previous chapter, we've added post creation, listing, update, and deletion functionality. This chapter will cover a comment's creation, listing, update, and deletion. We will create a page to display information about a post, add components to list comments, add a modal to display a form to create comments and add a dropdown to allow the user to delete or modify a comment. At the end of this chapter, you will learn how to navigate to a single page with URL parameters using React and React Router.

In this chapter, we will cover the following topics:

- Listing comments on a post page
- Creating a comment using a form
- Editing and deleting a comment
- Updating a comment

Technical requirements

Make sure to have VS Code and an updated browser installed and configured on your machine. You can find the code of this chapter at `https://github.com/PacktPublishing/Full-stack-Django-and-React/tree/chap9`.

Creating a UI

In the next paragraphs, we will modify the `Post` component for consistency when displaying a single post, add a **Back** button on the layout so the user can go back to the home page, and finally, add CRUD features, a little bit similar to the `Post` components. Before listing the comments, we need to ensure that the user can create comments. This will require building a page called `SinglePost` that will show details about a post and the comments.

Let's look at the UI of the page in the following figure:

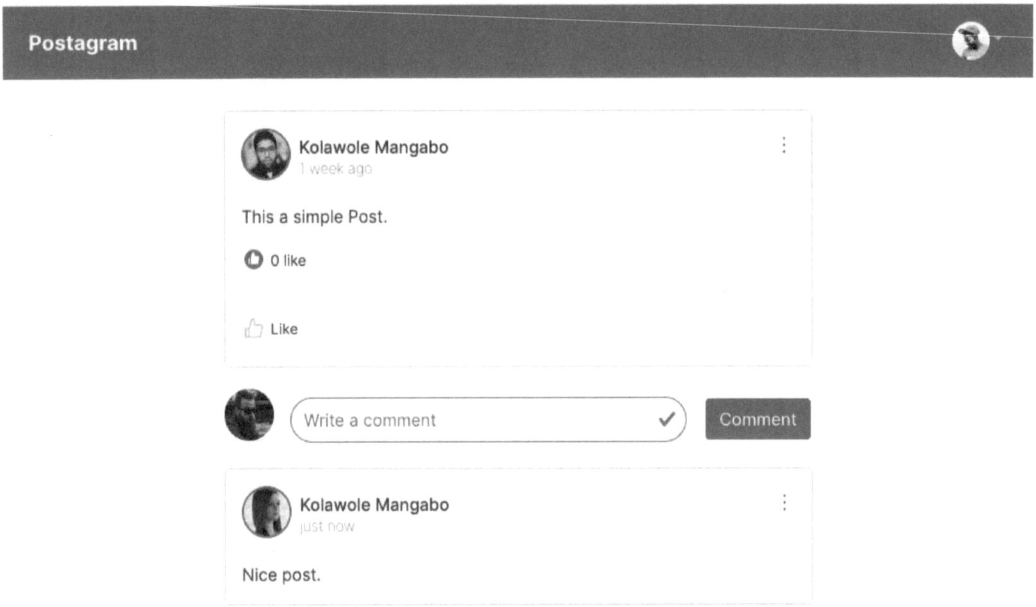

Figure 9.1 – Result of the SinglePost page

The UI in the preceding figure gives us a good idea of the result. When the page is built and the user clicks on a comment, a modal will appear, and the user will be able to create a comment. Let's stick to this case first, and we'll explore the other CRUD operations later.

Notice that we also have a back button in the top-left corner of the page – something to add to the `Layout` component. We will start by making some tweaks to the `Post.jsx` component first. This is because we are going to reuse the `Post` component, but we will mask options such as comment count and the **Comment** icon. After modifying the component, we will create a page displaying one article and comments.

Tweaking the Post component

The `Post` component will be simply reused to display more information about the post. Following the UI in *Figure 9.1*, we will just mask the number of comments on the post and the **Comment** icon.

Inside `Post.jsx`, we will add another prop called `isSinglePost`. When this prop is `true`, it means that we are showing the component on the `SinglePost` page:

src/components/posts/Post.jsx

```
...
function Post(props) {
```

```
const { post, refresh, isSinglePost } = props;
...
return (
  <>
   ...
          {!isSinglePost && (
           <p className="ms-1 fs-6">
             <small>
               <Link>
                 {post.comments_count} comments
               </Link>
             </small>
           </p>
          )}
   ...
              {!isSinglePost && (
          <div className="d-flex flex-row">
            <CommentOutlined
              style={{
                width: "24px",
                height: "24px",
                padding: "2px",
                fontSize: "20px",
                color: "#C4C4C4",
              }}
            />
            <p className="ms-1 mb-0">
              <small>Comment</small>
            </p>
          </div>
        )}
...
```

With the modification done to the `Post` component, we can now add the back button to the `Layout` component.

Adding a back button to the Layout component

The back button has the role of navigating the user to the preceding page if the action is initiated. An interesting idea about doing that is to add the actual path to the component where a go-back action can happen. However, it'll require a lot of code and will introduce some complexity.

Thankfully, the `react-router` library provides a simple way to navigate to the preceding page in just one line:

```
navigate(-1)
```

Yes! Let's add this function to the `Layout.jsx` component:

src/components/Layout.jsx

```
import { ArrowLeftOutlined } from "@ant-design/icons";
import { useNavigate } from "react-router-dom";

function Layout(props) {
 const { hasNavigationBack } = props;

 const navigate = useNavigate();
...
  return (
     <div>
       <Navigationbar />
       {hasNavigationBack && (
         <ArrowLeftOutlined
           style={{
             color: "#0D6EFD",
             fontSize: "24px",
             marginLeft: "5%",
             marginTop: "1%",
           }}
           onClick={() => navigate(-1)}
         />
       )}
       <div className="container my-2">
           {props.children}
```

```
        </div>
...
```

In the preceding code, we added a prop called `hasNavigationBack`. This prop will tell React whether it should render the icon to nagivate back to the precedent page. The rendering process is done in the JSX code, using conditional. If `hasNavigationBack` is `true`, we show the **Back** icon, and the user can navigate.

With the option of going back added, we can now move to write the `SinglePost.jsx` page.

Creating the SinglePost component

In the `src/pages` directory, create a new file called `SinglePost.jsx`. This file will contain the code to display information about a post and, most importantly, the comments. The following figure shows a simple wireframe of the page so we can have an idea about the layout of the components:

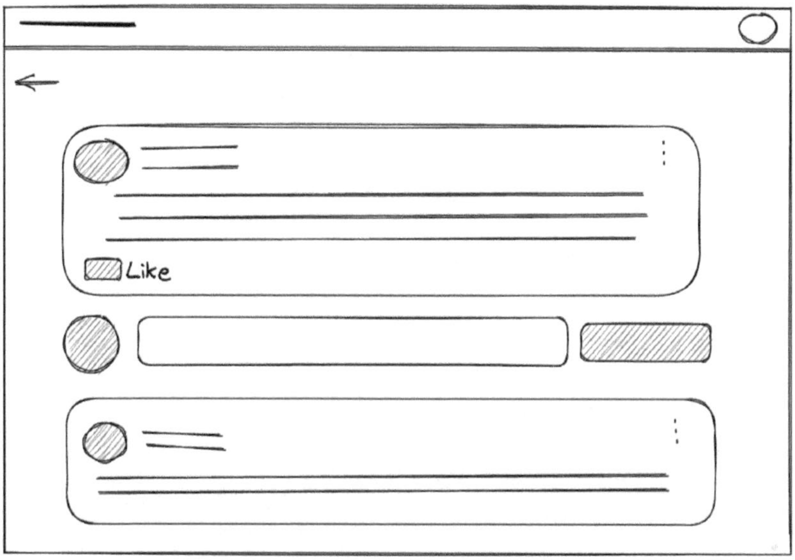

Figure 9.2 – Wireframe of the SinglePost page

Let's move the file and start coding. In the following snippet, we will create the `SinglePost` page, add the imports, and define the functions and states that will be used on the page:

src/pages/SinglePost.jsx

```
import React from "react";

import Layout from "../components/Layout";
```

```
import { Row, Col } from "react-bootstrap";
import { useParams } from "react-router-dom";
import useSWR from "swr";
import { fetcher } from "../helpers/axios";
import { Post } from "../components/posts";
import CreateComment from "../components/comments/
CreateComment";
import Comment from "../components/comments/Comment";

function SinglePost() {
 let { postId } = useParams();

 const post = useSWR(`/post/${postId}/`, fetcher);

 const comments = useSWR(`/post/${postId}/comment/`, fetcher);

 return (
   <Layout hasNavigationBack>
     {post.data ? (
       <Row className="justify-content-center">
         <Col sm={8}>
           <Post post={post.data} refresh={post.mutate}
             isSinglePost />
           // Adding CreateComment form and list all comments
              here
         </Col>
       </Row>
     ) : (
       <div>Loading...</div>
     )}
   </Layout>
 );
}

export default SinglePost;
```

We are using the Row and Col feature from react-bootstrap again. This structure will help us have one column taking 8/12 of the layout and having objects centered. Next, we need a form for comment creation.

We are also using a new Hook, useParams. As stated in the official documentation, the useParams Hook returns an object of **key/value** pairs of the dynamic params from the current URL that was matched by <Route path>. Child routes inherit all parameters from their parent routes.

A little bit complicated to grasp, but let's register this page and load it in the browser. Inside the App.jsx file, add a new route:

src/App.jsx

```
...
function App() {
  return (
    <Routes>
      <Route
        path="/"
        element={
          <ProtectedRoute>
            <Home />
          </ProtectedRoute>
        }
      />
      <Route
        path="/post/:postId/"
        element={
          <ProtectedRoute>
            <SinglePost />
          </ProtectedRoute>
        }
      />
...
```

The path of the newly added route has an interesting pattern with the addition of postId. We can tell react-router to expect a parameter that will be passed, and this parameter will then be available in the useParams Hook.

Let's add the redirection to the `SinglePost` page in the `Post` component:

src/components/posts/CreatePost.jsx

```
return (
    <>
    ...
            {!isSinglePost && (
              <p className="ms-1 fs-6">
                <small>
                  <Link to={`/post/${post.id}/`}>
                    {post.comments_count} comments
                  </Link>
                </small>
              </p>
            )}
    ...
```

Inside the `SinglePost.jsx` file, add a `console` log of `useParams()`:

src/pages/SinglePost.jsx

```
...
function SinglePost() {
  console.log(useParams())
  let { postId } = useParams();
...
```

Go into the browser and click on a post to access the `SinglePost` page. You will have a similar result:

Figure 9.3 – Post page

Check the browser console to see the content of `useParams()`:

Figure 9.4 – Content of useParams()

We have an object containing the `postId` value. With `useParams()` explained, let's move on to add the `CreateComment` form.

Creating a comment

Inside the `src/components` directory, create a new directory called `comments`. This directory will contain the code for the `comments` feature components. Inside the newly created directory, create a new file called `CreateComment.jsx`. This component represents the form that the user will use to add comments to a post.

Once the file is created, add the required imports:

src/components/comments/CreateComment.jsx

```
import React, { useState, useContext } from "react";
import { Button, Form, Image } from "react-bootstrap";
import axiosService from "../../helpers/axios";
import { getUser } from "../../hooks/user.actions";
import { randomAvatar } from "../../utils";
```

```
import { Context } from "../Layout";


function CreateComment(props) {
  const { postId, refresh } = props;

  return (
    <Form>

    </Form>
  );
}

export default CreateComment;
```

On the `CreateComment` page, we are going to show toast notifications when a CRUD action is made. That means that we are going to use the `Context` method again.

Let's start by defining the props and creating `handleSubmit`. This process will be pretty similar to what we've done in the `CreatePost` component:

src/components/comments/CreateComment

```
...
function CreateComment(props) {
  const { postId, refresh } = props;
  const [avatar, setAvatar] = useState(randomAvatar());
  const [validated, setValidated] = useState(false);
  const [form, setForm] = useState({});

  const { toaster, setToaster } = useContext(Context);

  const user = getUser();

  const handleSubmit = (event) => {
    // Logic to handle form submission
  };
...
```

Let's now add the `Form` UI:

src/component/comments/CreateComment.jsx

```
...
return (
   <Form
     className="d-flex flex-row justify-content-between"
     noValidate
     validated={validated}
     onSubmit={handleSubmit}
   >
     <Image
       src={avatar}
       roundedCircle
       width={48}
       height={48}
       className="my-2"
     />
     <Form.Group className="m-3 w-75">
       <Form.Control
         className="py-2 rounded-pill border-primary"
         type="text"
         placeholder="Write a comment"
         value={form.body}
         name="body"
         onChange={(e) => setForm({ ...form,
                                     body: e.target.value })}
       />
     </Form.Group>
     <div className="m-auto">
       <Button
         variant="primary"
         onClick={handleSubmit}
         disabled={form.body === undefined}
         size="small"
       >
```

```
          Comment
        </Button>
      </div>
    </Form>
  );
...
```

With the UI added, we can write the `handeSubmit` method:

src/component/comments/CreateComment.jsx

```
...
  const handleSubmit = (event) => {
    event.preventDefault();
    const createCommentForm = event.currentTarget;

    if (createCommentForm.checkValidity() === false) {
      event.stopPropagation();
    }
    setValidated(true);

    const data = {
      author: user.id,
      body: form.body,
      post: postId,
    };
    axiosService
      .post(`/post/${postId}/comment/`, data)
      .then(() => {
        setForm({ ...form, body: "" });
        setToaster({
          type: "success",
          message: "Comment posted successfully🚀",
          show: true,
          title: "Comment!",
        });
        refresh();
```

```
      })
      .catch(() => {
        setToaster({
          type: "danger",
          message: "",
          show: true,
          title: "An error occurred.!",
        });
      });
  };
...
```

Similar to the `CreatePost` component, we are doing checks on the validity of the form but also sending a request to the `/post/${postId}/comment/` endpoint. Then, depending on the response, we show a toast and clean the form. Let's test the form and add the first comment using React:

src/pages/SinglePost.jsx

```
...
  return (
    <Layout hasNavigationBack>
      {post.data ? (
        <Row className="justify-content-center">
          <Col sm={8}>
            <Post post={post.data} refresh={post.mutate}
              isSinglePost />
            <CreateComment postId={post.data.id}
              refresh={comments.mutate} />
          </Col>
        </Row>
      ) : (
        <div>Loading...</div>
      )}
    </Layout>
  );
...
```

You should have a similar result:

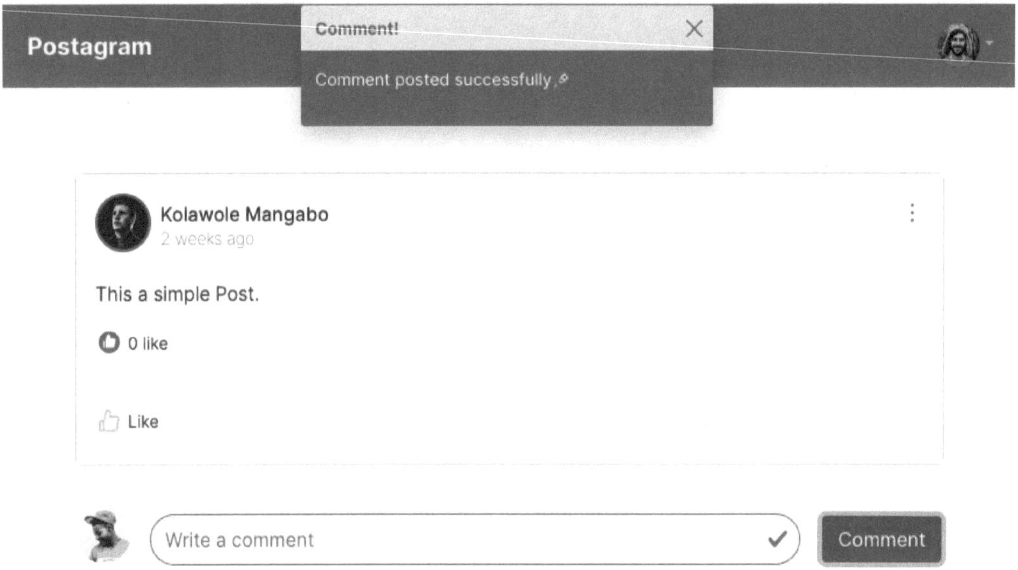

Figure 9.5 – Creating a comment

In the preceding paragraphs, we have created a page to display information about a post, thus allowing us to add a modal displaying a form to create a new comment related to this post.

Now, we need to display the created comments.

Listing the comments

We can create comments, but we can't see them. In `src/components/comments`, create a new file called `CreateComment.jsx`. This will contain the code for the `Comment` component that will be used to show details about a comment. Here's a wireframe of the `Comment` component:

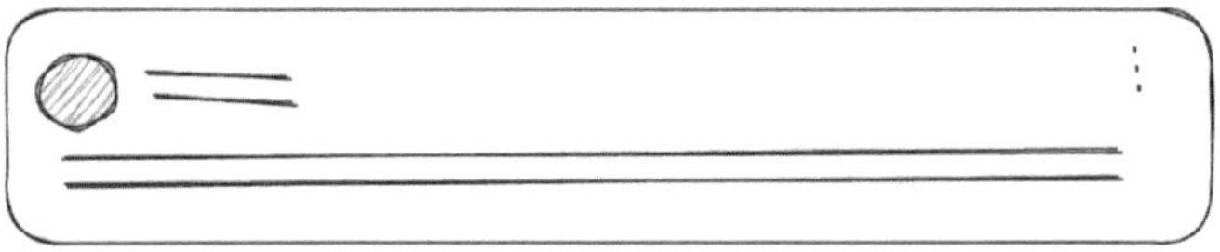

Figure 9.6 – Wireframe of the Comment component

Let's move on to writing the code. Let's start by adding the `CreateComment` function and the imports, and defining the state that we will use in this component:

src/components/comments/CreateComment.jsx

```
import React, { useState, useContext } from "react";
import { format } from "timeago.js";
import { Image, Card, Dropdown } from "react-bootstrap";
import { randomAvatar } from "../../utils";
import axiosService from "../../helpers/axios";
import { getUser } from "../../hooks/user.actions";
import UpdateComment from "./UpdateComment";
import { Context } from "../Layout";
import MoreToggleIcon from "../MoreToggleIcon";

function Comment(props) {
 const { postId, comment, refresh } = props;
 const { toaster, setToaster } = useContext(Context);

 const user = getUser();

 const handleDelete = () => {
    // Handle the deletion of a comment
 };

 return (
   <Card className="rounded-3 my-2">
     // Code for the comment card
   </Card>
 );
}

export default Comment;
```

We have the necessary imports. Let's start with the UI first. It's a little bit like the `Post` component in its structure:

src/components/comments/CreateComment.jsx

```
...
return (
  <Card className="rounded-3 my-2">
    <Card.Body>
      <Card.Title className="d-flex flex-row
        justify-content-between">
        <div className="d-flex flex-row">
          <Image
            src={randomAvatar()}
            roundedCircle
            width={48}
            height={48}
            className="me-2 border border-primary
                       border-2"
          />
          <div className="d-flex flex-column
              justify-content-start
              align-self-center mt-2">
            <p className="fs-6 m-0">{comment.author.name}
            </p>
            <p className="fs-6 fw-lighter">
              <small>{format(comment.created)}</small>
            </p>
          </div>
        </div>
        {user.name === comment.author.name && (
          <div>
            <Dropdown>
              <Dropdown.Toggle
                as={MoreToggleIcon}></Dropdown.Toggle>
              <Dropdown.Menu>
                <Dropdown.Item>
```

```
                Modify
              </Dropdown.Item>
              <Dropdown.Item onClick={handleDelete}
                className="text-danger">
                Delete
              </Dropdown.Item>
            </Dropdown.Menu>
          </Dropdown>
        </div>
      )}
    </Card.Title>
    <Card.Text>{comment.body}</Card.Text>
  </Card.Body>
</Card>
);
...
```

The UI for the `Comment` component is ready. Let's see the result on a post page:

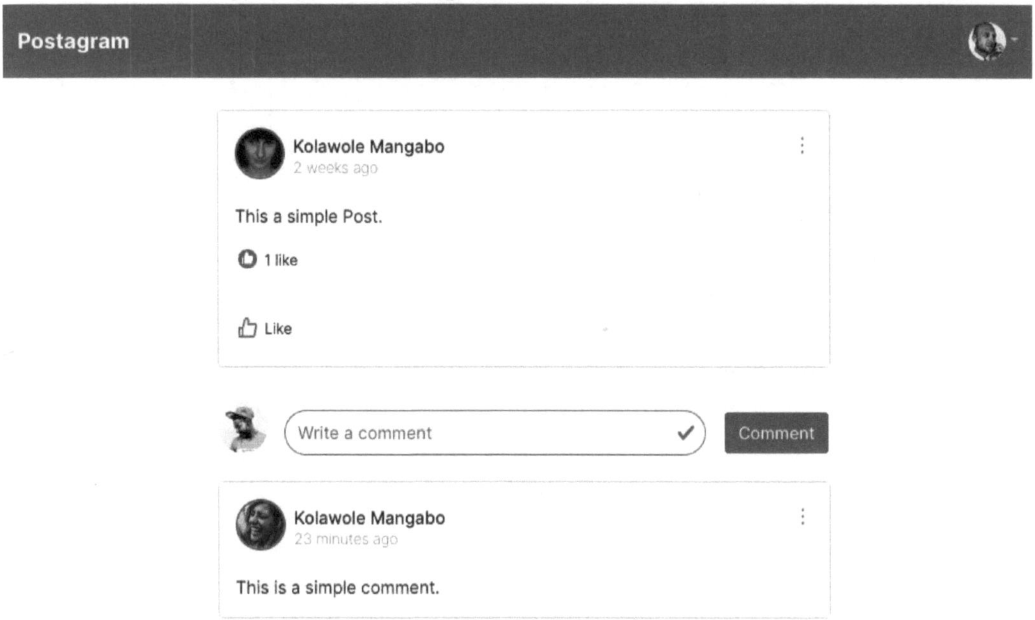

Figure 9.7 – List of comments on a post

We also have the **More** dots in the top-right corner of each component, meaning that we need to implement features for the deletion and modification of a comment. Let's add the deletion feature.

Deleting a comment

The **More** dots menu provides two options: deleting and modifying a comment. Let's start by adding code and actions to delete a comment. The function has already been declared; we just need to add the logic:

Src/components/comments/CreateComment.jsx

```
...
  const handleDelete = () => {
    axiosService
      .delete(`/post/${postId}/comment/${comment.id}/`)
      .then(() => {
        setToaster({
          type: "danger",
          message: "Comment deleted 🚀",
          show: true,
          title: "Comment Deleted",
        });
        refresh();
      })
      .catch((err) => {
        setToaster({
          type: "warning",
          message: "Comment deleted 🚀",
          show: true,
          title: "Comment Deleted",
        });
      });
  };
...
```

In the `handleDelete` function, we make a request using `axios` to `/post/${postId}/comment/${comment.id}/` to delete a comment. Depending on the result of the HTTP request, we show a toaster with the correct message. Once you are done adding the code, let's test the result:

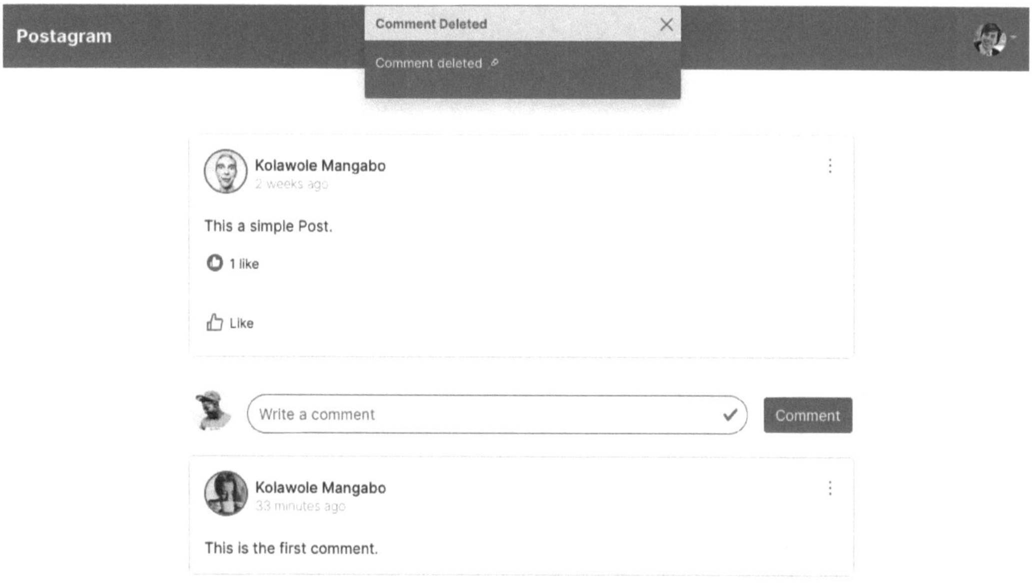

Figure 9.8 – Deleting a comment

The deletion of a comment in our React application is now possible. Let's move on to adding the feature for modifying comments.

Updating a comment

Updating a comment will be similar to what was done in the `UpdatePost.jsx` file. However, I'll assist you in writing this feature for the comments. We also have an exciting element to add to our comments: liking and unliking a comment, but as an exercise. Let's focus on the modification of a comment. For this purpose, we will have to create a modal.

Adding the UpdateComment modal

Inside the `src/components/comments` directory, create a file called `UpdateComment.jsx`. This file will contain the modal and the form that will allow the user to update a comment:

src/components/comments/UpdateComment.jsx

```
import React, { useState, useContext } from "react";
import { Button, Modal, Form, Dropdown } from "react-
bootstrap";
import axiosService from "../../helpers/axios";
```

```
import { Context } from "../Layout";


function UpdateComment(props) {
 const { postId, comment, refresh } = props;
 const [show, setShow] = useState(false);
 const [validated, setValidated] = useState(false);
 const [form, setForm] = useState({
   author: comment.author.id,
   body: comment.body,
   post: postId
 });

 const { toaster, setToaster } = useContext(Context);

 const handleSubmit = (event) => {
   // handle the modification of a comment
 };

 return (
   <>
     <Dropdown.Item
       onClick={handleShow}>Modify</Dropdown.Item>

     // Adding the Modal here
   </>
 );
}

export default UpdateComment;
```

We are doing the required imports and defining the states that will be used and updated when a modification is triggered. Note that we also pass `postId` and the `comment` object as **props**. The first is needed for the endpoint; the second is also for the endpoint, but most importantly, to have a default value, we need to show it in the form for the user to modify.

Let's add the modal UI:

src/components/comments/UpdateComment.jsx

```
...
  return (
    <>
      <Dropdown.Item onClick={handleShow}>Modify
      </Dropdown.Item>

      <Modal show={show} onHide={handleClose}>
        <Modal.Header closeButton className="border-0">
          <Modal.Title>Update Post</Modal.Title>
        </Modal.Header>
        <Modal.Body className="border-0">
          <Form noValidate validated={validated}
            onSubmit={handleSubmit}>
            <Form.Group className="mb-3">
              <Form.Control
                name="body"
                value={form.body}
                onChange={(e) => setForm({ ...form,
                  body: e.target.value })}
                as="textarea"
                rows={3}
              />
            </Form.Group>
          </Form>
        </Modal.Body>
        <Modal.Footer>
          <Button variant="primary" onClick={handleSubmit}>
            Modify
          </Button>
        </Modal.Footer>
      </Modal>
    </>
  );
...
```

With the UI ready, we can now write the `handleSubmit` function:

src/components/comments/UpdateComment.jsx

```
...
const handleSubmit = (event) => {
  event.preventDefault();
  const updateCommentForm = event.currentTarget;

  if (updateCommentForm.checkValidity() === false) {
    event.stopPropagation();
  }

  setValidated(true);

  const data = {
    author: form.author,
    body: form.body,
    post: postId
  };

  axiosService
    .put(`/post/${postId}/comment/${comment.id}/`, data)
    .then(() => {
      handleClose();
      setToaster({
        type: "success",
        message: "Comment updated 🚀",
        show: true,
        title: "Success!",
      });
      refresh();
    })
    .catch((error) => {
      setToaster({
        type: "danger",
        message: "An error occurred.",
```

```
        show: true,
        title: "Comment Error",
      });
    });
  };
...
```

Let's import and add this component to the `Comment.jsx` file:

src/components/comments/Comment.jsx

```
...
        {user.name === comment.author.name && (
          <div>
            <Dropdown>
              <Dropdown.Toggle as={MoreToggleIcon}>
              </Dropdown.Toggle>
              <Dropdown.Menu>
                <UpdateComment
                  comment={comment}
                  refresh={refresh}
                  postId={postId}
                />
                <Dropdown.Item onClick={handleDelete}
                  className="text-danger">
                  Delete
                </Dropdown.Item>
              </Dropdown.Menu>
            </Dropdown>
          </div>
        )}
      </Card.Title>
      <Card.Text>{comment.body}</Card.Text>
    </Card.Body>
  </Card>

...
```

After adding this piece of code, once you click on the **Modify** option of the **More** menu, a modal will appear, like in the following figure:

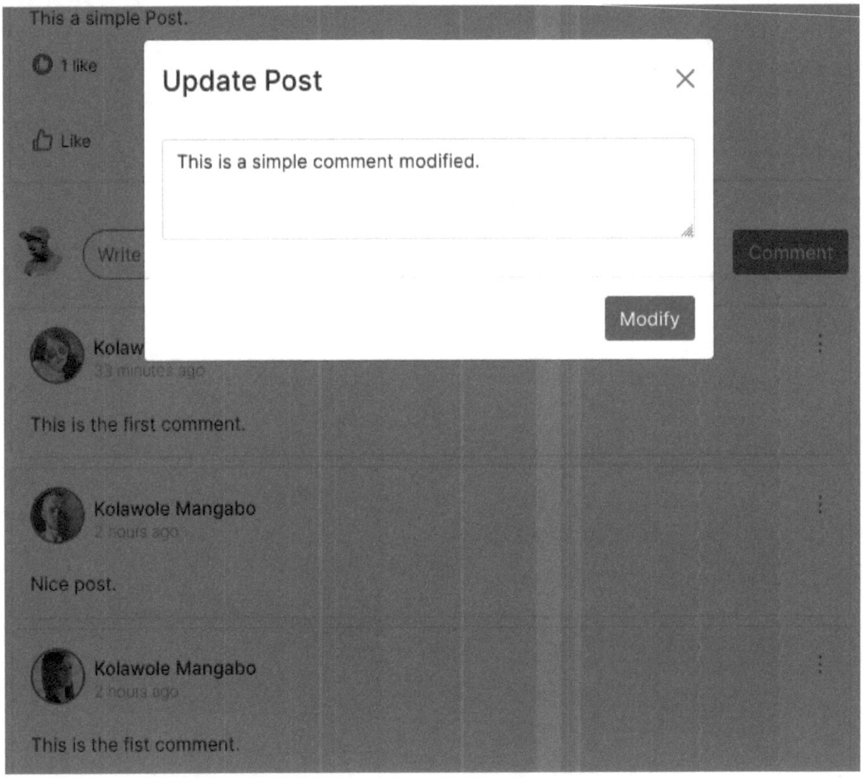

Figure 9.9 – Modify comment modal

If the modification is submitted and successful, a toast will appear at the top of the page:

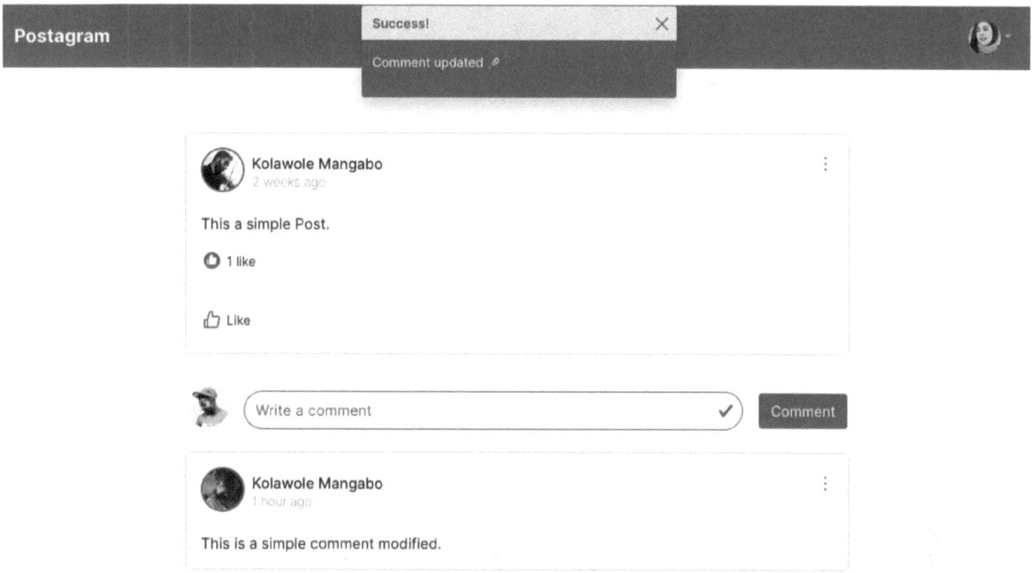

Figure 9.10 – Toast showing successful comment modification

Nice! We have completed working on CRUD actions for the comment feature. An exciting feature to have for the comments is the possibility to like a comment. It is similar to what we have done for the posts. This is the next step for this chapter, but also an exercise.

Liking a comment

Adding the **Like** feature to the **Comment** feature will require some changes to the Django API and some code to be added to the React application. First, let me provide you with the final result:

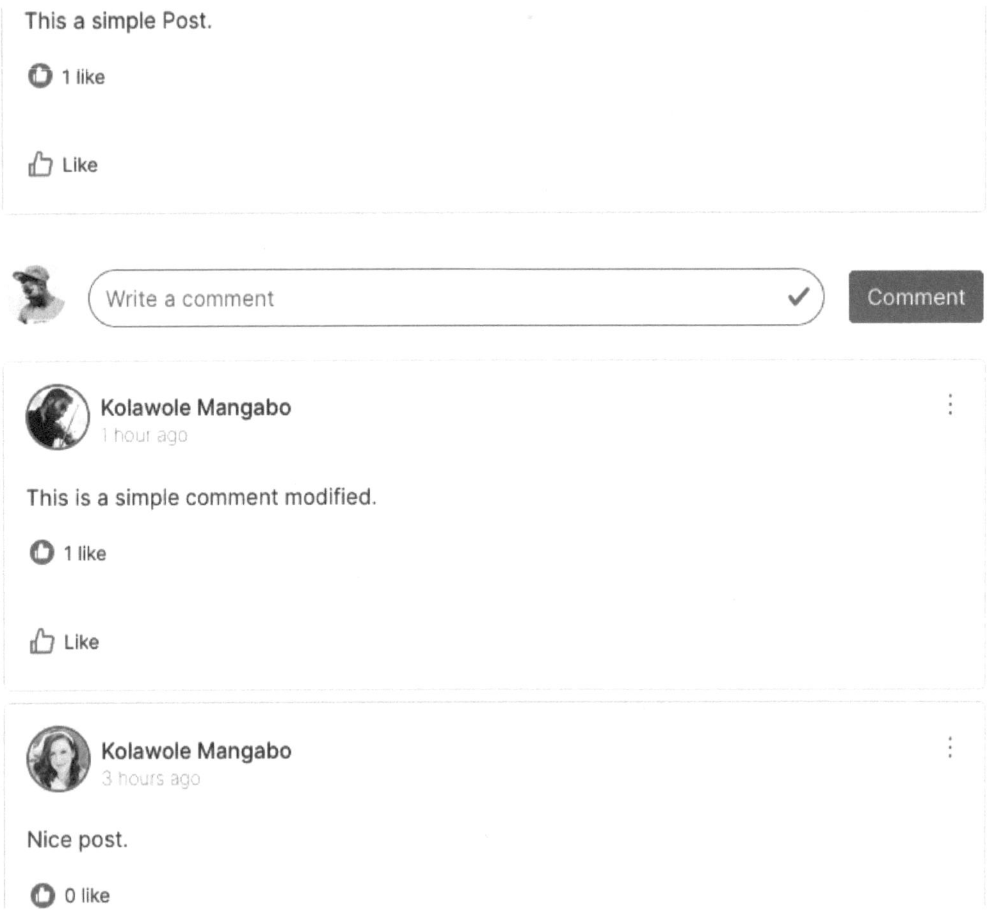

Figure 9.11 – Comments with the Like feature and likes count

Here is a list of the requirements of the feature:

- The user can see the number of likes on a comment
- The user can like a comment
- The user can remove a like from a comment

This will require some tweaks on the Django API as well. Feel free to get inspired by what we've done for the post feature.

Good luck with the exercise. You can find the solution at `https://github.com/PacktPublishing/Full-stack-Django-and-React/blob/main/social-media-react/src/components/comments/Comment.jsx`.

After adding the **Like** feature to comments, we are now ready to finally add CRUD operations to the profile of the React application. We will create a profile page and allow the user to edit the information in their profile. We will also enable the user to update their avatar and set a default avatar image for users.

Summary

In this chapter, we focused on adding CRUD operations to the comments feature. We've learned how to play with `react-router` Hooks to retrieve parameters and use them in the code. We've also added a **Like** feature to the comment. A user can like or unlike a comment, and also see the number of likes for a post. That led us to learn more about the `useState` and `useContext` Hooks and the way they affect a state in React. We have also learned how to create a dropdown component, how to use the custom toaster, and how to tweak a component to fits some requirements.

In the next chapter, we'll focus on CRUD operations on the user profile, and we will also learn how to upload a profile picture.

Questions

1. What is the usage of `useParams`?
2. How do you write a route in React that can support parameter passing?
3. What is the use of the `useContext` Hook?

10
User Profiles

A social media application should allow users to consult other user profiles. From another view, it should also allow an authenticated user to edit their information, such as their last name, first name, and avatar.

In this chapter, we will focus on adding CRUD features on the user side. We'll build a page to visualize a user profile and a page that allows a user to edit their information. This chapter will cover the following topics:

- Listing profiles on the home page
- Displaying user information on their profile page
- Editing user information

Technical requirements

Make sure to have VS Code and an updated browser installed and configured on your machine. You can find all the code files used in this chapter at `https://github.com/PacktPublishing/Full-stack-Django-and-React/tree/chap10`.

Listing profiles on the home page

Before building the pages and components to display user information and allow user information modification, we need to add a component to list some profiles on the home page like so:

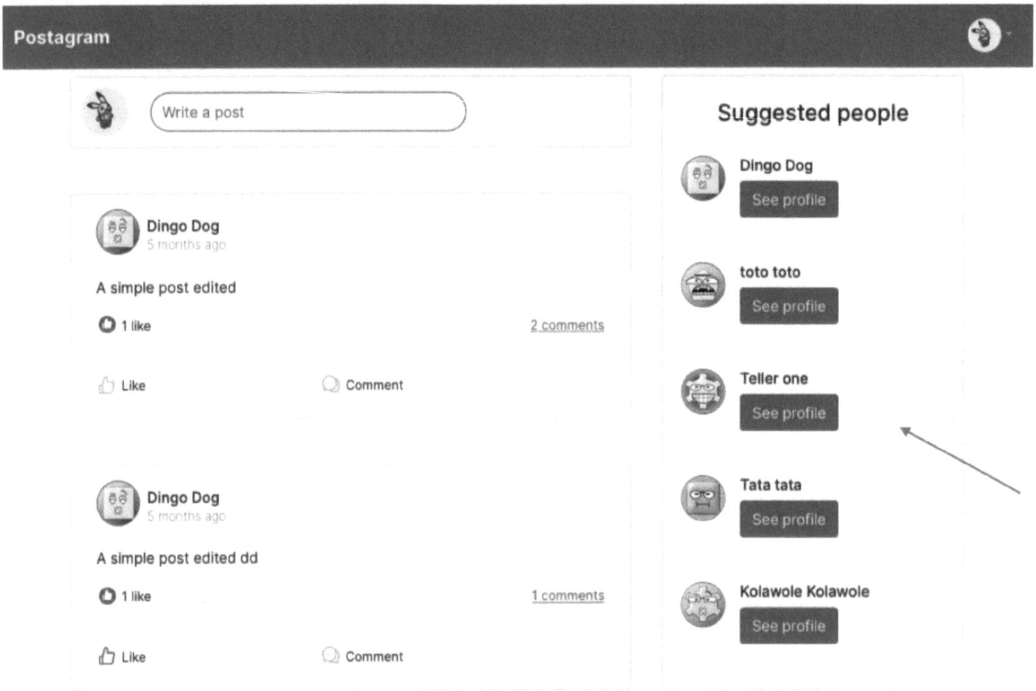

Figure 10.1 – Listing profiles

Follow these steps to add the component for listing profiles on the home page:

1. In the `src/components` file, create a new directory called `profile`. This directory will contain the code for all components related to users or profiles.
2. In the newly created directory, create a file called `ProfileCard.jsx` and add the following content:

src/components/profile/ProfileCard.jsx

```
import React from "react";
import { Card, Button, Image } from "react-bootstrap";
import { useNavigate } from "react-router-dom";

function ProfileCard(props) {

 return (
    // JSX code here
 );
```

```
}

export default ProfileCard;
```

The `ProfileCard` component will be used to display information about a profile and redirect the user to the profile page.

3. Next, we will add the code logic concerning the navigation to the profile page and the props object destructuration:

src/components/profile/ProfileCard.jsx

```
...
function ProfileCard(props) {
  const navigate = useNavigate();
  const { user } = props;

  const handleNavigateToProfile = () => {
    navigate(`/profile/${user.id}/`)
  };

  return (
     // JSX Code
  );
}

export default ProfileCard;
```

In the preceding code, we retrieved the user object from the props and we also added a function to handle the navigation to the user profile.

4. Next, let's write the JSX that will display information to the user:

src/components/profile/ProfileCard.jsx

```
...
  return (
    <Card className="border-0 p-2">
      <div className="d-flex ">
        <Image
          src={user.avatar}
```

```
          roundedCircle
          width={48}
          height={48}
          className=
            "my-3 border border-primary border-2"
        />
        <Card.Body>
          <Card.Title
            className="fs-6">{user.name}</Card.Title>
          <Button variant="primary"
            onClick={handleNavigateToProfile}>
            See profile
          </Button>
        </Card.Body>
      </div>
    </Card>
  );
}

export default ProfileCard;
```

The `ProfileCard` component is written. We can now import it into the `Home.jsx` page and use it. But before that, we need to retrieve five profiles from the API and loop through the results to have the wanted display:

src/pages/Home.jsx

```
...
import CreatePost from "../components/posts/CreatePost";
import ProfileCard from "../components/profile/
ProfileCard";

function Home() {
...
 const profiles = useSWR("/user/?limit=5", fetcher);
...

 return (
```

```
    <Layout>
      <Row className="justify-content-evenly">
        ...
        <Col sm={3} className="border rounded py-4
            h-50">
          <h4 className="font-weight-bold text-center">
            Suggested people</h4>
          <div className="d-flex flex-column">
            {profiles.data &&
              profiles.data.results.map((profile,
                                         index) => (
                <ProfileCard key={index} user={profile}
                />
              ))}
          </div>
        </Col>
      </Row>
    </Layout>
  );
}

export default Home;
```

In the preceding code, the profiles are only shown if the `profiles.data` object is not null or undefined. This is why we are writing the `profiles.data && profiles.data.results.map()` inline JSX condition.

5. Once it's done, reload the home page and you'll have a new component available, listing a maximum of five profiles.

   Try to click on the **See Profile** button. You will be redirected to a white page. This is normal because we haven't written routing for the **Profile** page yet.

In the next section, we will be creating components to display information about a profile, like so:

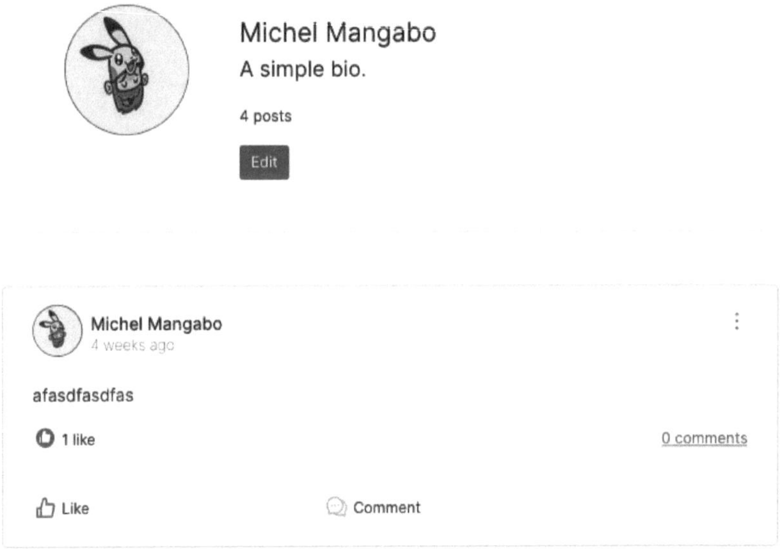

Figure 10.2 – The user profile page

We will also allow the user to edit their information, like so:

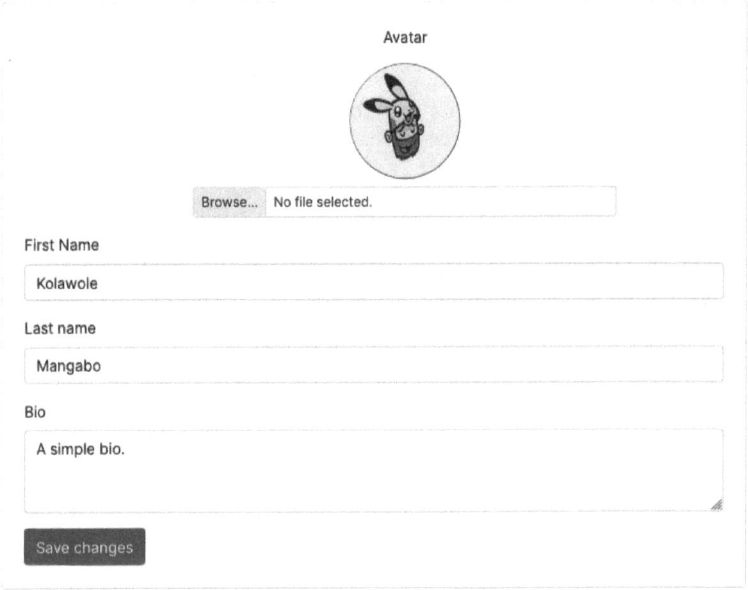

Figure 10.3 – The user edit form and page

Displaying user information on their profile page

In this section, we will create a profile page to display user information. We will build a component to display user details and the posts concerning this user, but also we will create a page displaying a form for editing user information.

Before starting to build the user profile page, we have to create some components. On the profile page, we are not only displaying information but also the list of posts created by the user. Let's start by writing the `ProfileDetails.jsx` component (*Figure 10.4*):

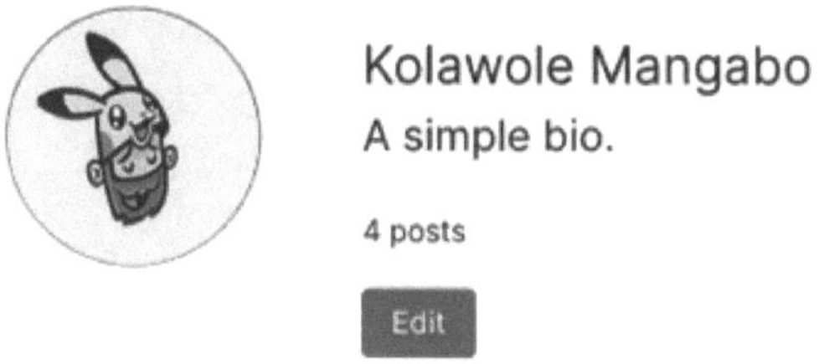

Figure 10.4 – The ProfileDetails component

Here's the wireframe to help you with the structure of the component:

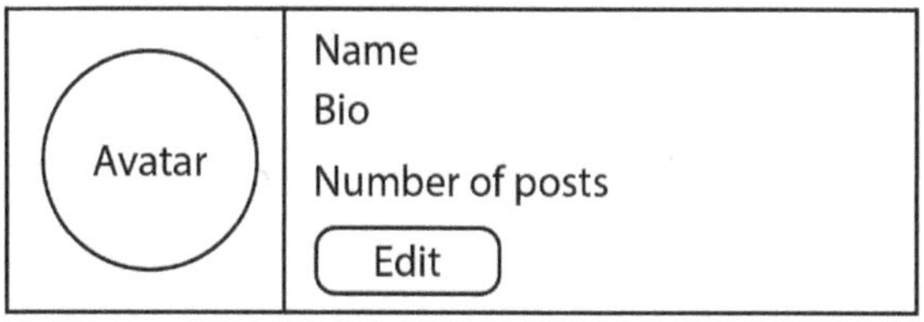

Figure 10.5 – The wireframe of the ProfileDetails component

In the `ProfileDetails` component, we are displaying some avatars. At this point of the project, it's time to get rid of the `randomAvatar` function. It has been useful up until this point of the project, but we are making a lot of requests and some state change within the application just calls the function again that returns another random image, which is not something a user of the application might want to see.

Let's start using the value of the avatar field on the user object but before that, we have to configure Django to deal with media upload and the avatar field on the user object.

The social media application uses the `avatar` field, which represents a link to a file that the browser can make a request on and receive the image. Django supports file uploading; we just need to add some configuration to make it effective.

Inside the `settings.py` file of the project, add the following lines at the end of the project:

CoreRoot/settings.py

```
...
MEDIA_URL'= '/med'a/'
MEDIA_ROOT = BASE_DIR'/ 'uplo'ds'
```

The `MEDIA_URL` setting allows us to write the URL that will be used to retrieve uploaded files. The `MEDIA_ROOT` setting tells Django where to stock the files and also checks the upload files when returning the URL of a file. In the case of this project, an avatar field will have this URL, for example: `http://localhost:8000/media/user_8380ca50-ad0f-4141-88ef-69dc9b0707ad/avatar-rogemon.png`.

For this configuration to work, you will need to create a directory called `uploads` at the root of the Django project. You will also need to install the Pillow library, which contains all the basic tools for image processing functionality:

```
pip install pillow
```

After that, let's slightly modify the avatar field on the user model. Inside the `core/user/models.py`, add a function before the `UserManager` manager class:

core/user/models.py

```
...
def user_directory_path(instance, filename):
    # file will be uploaded to
        MEDIA_ROOT/user_<id>/<filename>
    return 'user_{0}/{1}'.format(instance.public_id,
                                 filename)
...
```

This function will help re-write the path for the upload of a file. Instead of going directly into the `uploads` directory, the avatar is stocked according to a user. It can help with the better organization of files in your system. After adding the function, we can tell Django to use it for the default upload path:

core/user/models.py

```
...
class User(AbstractModel, AbstractBaseUser, PermissionsMixin):
```

```
...
    avatar = models.ImageField(
        null=True, blank=True,
            upload_to=user_directory_path)
...
```

In Django, the `ImageField` field is used to store image files in a database. It is a subclass of `FileField`, which is a generic field for storing files, so it has all the attributes of `FileField` as well as some additional attributes specific to images. The `upload_to` attribute specifies the directory where the image files will be stored.

Now, run the `makemigrations` command and make sure to migrate the changes to the database:

```
python manage.py makemigrations
python manage.py migrate
```

With this configuration done, our API can accept avatar uploading for the user. However, some users won't have an avatar and we have been handling it pretty badly from the frontend side. Let's set up a default avatar that will be used for users without an avatar.

Configuring the default avatar

To configure the default avatar, follow these steps:

1. In the `settings.py` file of the Django project, add the following line at the end of the file:

CoreRoot/settings.py

```
...
DEFAULT_AVATAR_URL = "https://avatars.dicebear.com/api/
identicon/.svg"
```

The avatar image looks as follows:

Image 10.6: The default image

2. Once you have added DEFAULT_AVATAR_URL to the settings.py file, we will slightly modify the UserSerializer representation method to return the DEFAULT_AVATAR_URL value by default if the avatar field is none:

Core/user/serializers.py

```
...
from django.conf import settings


class UserSerializer(AbstractSerializer):
...
    def to_representation(self, instance):
        representation =
          super().to_representation(instance)
        if not representation['avatar']:
            representation['avatar'] =
              settings.DEFAULT_AUTO_FIELD
            return representation
        if settings.DEBUG:  # debug enabled for dev
            request = self.context.get('request')
            representation['avatar'] =
              request.build_absolute_uri(
                representation['avatar'])
        return representation
```

Let's explain what we are doing in the preceding code block. First, we need to check whether the avatar value exists. If that's not the case, we will return the default avatar. By default, Django doesn't return the actual route of the file with the domain. That's why in this case, if we are in a development environment, we return an absolute URL of the avatar. In the last part of this book, we will deploy the application on a production server, then we will use **AWS S3** for file storing.

With the fix done on the backend, we can confidently modify the frontend application by now including the avatar field. It's quite simple and a little bit of refactoring. Remove the randomAvatar function code from the React application and replace the values with user.avatar, post.author.avatar, or comment.author.avatar, depending on the file and the component.

3. With those small configurations done, check the **Home** page; you should have a similar result.

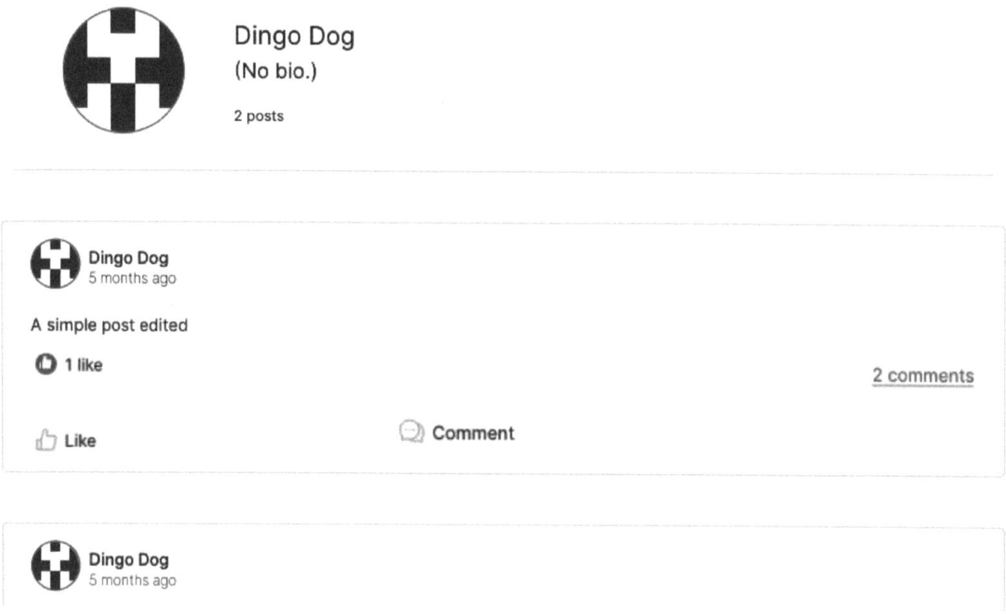

Figure 10.7 – The Home page with a default avatar

Great! Let's move to creating the **Profile** page so that our Django application is ready to accept file uploads.

Writing the ProfileDetails component

To create the `ProfileDetails` component, we have to create the file that will contain the code for this component, add the logic for the navigation, write the UI (JSX), and import the component on the **Profile** page:

1. In the `src/components/profile` directory, create a new file called `ProfileDetail.jsx`. This file will contain the code for the `ProfileDetails` component:

src/components/profile/ProfileDetails.jsx

```
import React from "react";
import { Button, Image } from "react-bootstrap";
import { useNavigate } from "react-router-dom";

function ProfileDetails(props) {

 return (
```

```
    // JSX code here
  );
}

export default ProfileDetails;
```

2. Here, we just need to destruct the props object to retrieve the user object, declare the navigate variable to use the `useNagivate` Hook, and finally handle the case when the user object is undefined or null:

src/components/profile/ProfileDetaisl.jsx

```
...
function ProfileDetails(props) {
  const { user } = props;
  const navigate = useNavigate();

  if (!user) {
    return <div>Loading...</div>;
  }

  return (
    // JSX Code here
  );
}

export default ProfileDetails;
```

3. We can confidently write the JSX logic now:

src/components/profile/ProfileDetails.jsx

```
...
  return (
    <div>
      <div className="d-flex flex-row border-bottom
        p-5">
        <Image
          src={user.avatar}
```

```
          roundedCircle
          width={120}
          height={120}
          className="me-5 border border-primary
                     border-2"
        />
        <div className="d-flex flex-column
         justify-content-start align-self-center mt-2">
          <p className="fs-4 m-0">{user.name}</p>
          <p className="fs-5">{user.bio ? user.bio :
            "(No bio.)"}</p>
          <p className="fs-6">
            <small>{user.posts_count} posts</small>
          </p>
          <Button
            variant="primary"
            size="sm"
            className="w-25"
            onClick={() =>
              navigate(`/profile/${user.id}/edit/`)}
          >
            Edit
          </Button>
        </div>
      </div>
    </div>
  );
```

4. Now that the component is written, create a new file called `Profile.jsx` in the `src/pages` directory. This file will contain the code and logic for the **Profile** page:

src/pages/Profile.jsx

```
import React from "react";
import { useParams } from "react-router-dom";
import Layout from "../components/Layout";
import ProfileDetails from "../components/profile/
ProfileDetails";
```

```
import useSWR from "swr";
import { fetcher } from "../helpers/axios";
import { Post } from "../components/posts";
import { Row, Col } from "react-bootstrap";

function Profile() {


  return (
    // JSX CODE
  );
}


export default Profile;
```

5. Let's add the fetching logic for the user and the user posts. No need to create another `Post` component as the same `Post` component from `src/components/Post.jsx` will be used to list posts created by the profile:

src/pages/Profile.jsx

```
...
function Profile() {
  const { profileId } = useParams();


  const user = useSWR(`/user/${profileId}/`, fetcher);


  const posts = useSWR(`/post/?author__public_
id=${profileId}`, fetcher, {
        refreshInterval: 20000
    });
...
```

6. Once it's done, we can now write the UI logic:

src/pages/Profile.jsx

```
...
  return (
    <Layout hasNavigationBack>
      <Row className="justify-content-evenly">
        <Col sm={9}>
          <ProfileDetails user={user.data}/>
          <div>
            <Row className="my-4">
              {posts.data?.results.map((post, index)
                => (
                <Post key={index} post={post}
                  refresh={posts.mutate} />
              ))}
            </Row>
          </div>
        </Col>
      </Row>
    </Layout>
  );
}
...
```

7. Great! Let's now register this page in the `App.js` file:

src/App.js

```
...
<Route
  path="/profile/:profileId/"
  element={
    <ProtectedRoute>
      <Profile />
    </ProtectedRoute>
  }
/>
...
```

8. Let's not forget to add `Link` to the profile in the `Navbar.jsx` file:

src/components/Navbar.jsx

```
...
      <NavDropdown.Item as={Link} to=
        {`/profile/${user.id}/`}>Profile
      </NavDropdown.Item>
      <NavDropdown.Item onClick={handleLogout}>Logout
      </NavDropdown.Item>
    </NavDropdown>
  </Nav>
...
```

9. Great! You can now click on the **See Profile** button or directly on the drop-down menu of the navigation bar to go to the profile page:

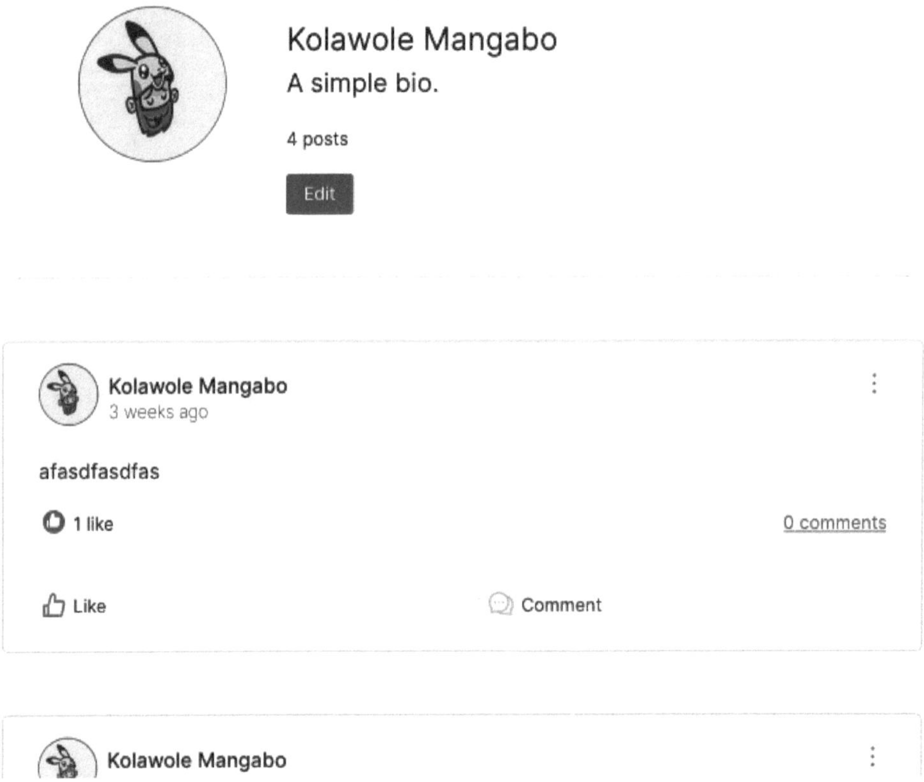

Figure 10.8 – A random profile page

With the profile page ready, we can move on to create the page that will contain the form to edit user information.

Editing user information

The **Edit** button on the **Profile** page should redirect the user to a page with a form to update its information. In the end, you will have a similar result to that shown in *Figure 10.2*. In this section, we will modify the `useUserActions` Hook by adding a new method to edit user information via the API. Then, we will create the form to edit user information. Lastly, we will integrate the editing form component on the `EditUser` page.

Let's start by adding a new method to the `useUserActions` Hook.

Adding the edit method to useUserActions

In the `src/hooks/user.actions.js` file, we will add another method to the `useUserActions` Hook. This function will handle the `patch` request to the API. As we are saving a user object in `localStorage`, we will update the value of the object if the request succeeds:

src/hooks/user.actions.js

```
function useUserActions() {
  const navigate = useNavigate();
  const baseURL = "http://localhost:8000/api";
  return {
    login,
    register,
    logout,
    edit
  };
...
  // Edit the user
  function edit(data, userId) {
    return axiosService.patch(`${baseURL}/user/${userId}/`,
                             data).then((res) => {
      // Registering the account in the store
      localStorage.setItem(
        "auth",
        JSON.stringify({
          access: getAccessToken(),
          refresh: getRefreshToken(),
```

```
            user: res.data,
          })
        );
      });
    }
  ...
}
```

With the `edit` function written, we can confidently move to create the form used to edit user information.

The UpdateProfileForm component

In the `src/components/UpdateProfileForm.jsx`, create a file called `UpdateProfileForm.jsx`. This file will contain the code for the component used to edit user information:

src/components/UpdateProfileForm.jsx

```
import React, { useState, useContext } from "react";
import { Form, Button, Image } from "react-bootstrap";
import { useNavigate } from "react-router-dom";

import { useUserActions } from "../../hooks/user.actions";
import { Context } from "../Layout";

function UpdateProfileForm(props) {
  return (
    // JSX Code
  );
}

export default UpdateProfileForm;
```

Let's start by retrieving the user object from the props and adding the Hooks needed for the form handling:

src/components/UpdateProfileForm.jsx

```
...
function UpdateProfileForm(props) {
  const { profile } = props;
```

```
  const navigate = useNavigate();

  const [validated, setValidated] = useState(false);
  const [form, setForm] = useState(profile);
  const [error, setError] = useState(null);
  const userActions = useUserActions();

  const [avatar, setAvatar] = useState();

  const { toaster, setToaster } = useContext(Context);
...
```

The next step is to write the `handleSubmit` method. This method should handle the validity of the form, the request to update the information, and also what to display according to the result:

src/components/UpdateProfileForm.jsx

```
...
const handleSubmit = (event) => {
  event.preventDefault();
  const updateProfileForm = event.currentTarget;


  if (updateProfileForm.checkValidity() === false) {
    event.stopPropagation();
  }
  setValidated(true);

  const data = {
    first_name: form.first_name,
    last_name: form.last_name,
    bio: form.bio,
  };

  const formData = new FormData();
}
...
```

As we are going to include a file in the data sent to the server, we are using a `FormData` object. A `FormData` object is a common way to create a bundle of data that will be sent to a server. It provides a simple and easy way to construct a set of **key/value** pairs, representing the name of the form fields and their value.

In the case of our project, we will need to pass the data in the data variable to the `formData` object:

src/components/UpdateProfileForm.jsx

```
...
const formData = new FormData();

Object.keys(data).forEach((key) => {
    if (data[key]) {
      formData.append(key, data[key]);
    }
});
...
```

The `Object` constructor provides a `keys` method that returns the list of keys in a JavaScript object. We then use the `forEach` method to loop through the `keys` array, check if `data [key]` value is not null, and then we append the values from the data object to the `formData` object. We also need to add a case for the avatar field:

src/components/UpdateProfileForm.jsx

```
...
const formData = new FormData();


// Checking for null values in the form and removing them.


Object.keys(data).forEach((key) => {
    if (data[key]) {
      formData.append(key, data[key]);
    }
});
```

```
if (avatar) {
  formData.append("avatar", avatar);
}
...
```

We can now move to the edit action:

src/components/UpdateProfileForm.jsx

```
...
userActions
  .edit(formData, profile.id)
  .then(() => {
    setToaster({
      type: "success",
      message: "Profile updated successfully 🚀",
      show: true,
      title: "Profile updated",
    });
    navigate(-1);
  })
  .catch((err) => {
    if (err.message) {
      setError(err.request.response);
    }
  });
...
```

Nothing complicated here. It's like what we used to do for other requests on the API. Let's move to the form now. The form will contain fields for the avatar such as the first name, the last name, and the bio. These fields are the only information the user will update. Let's start by writing the avatar field first:

src/components/UpdateProfileForm.jsx

```
...
return (
  <Form
    id="registration-form"
```

```
    className="border p-4 rounded"
    noValidate
    validated={validated}
    onSubmit={handleSubmit}
  >
    <Form.Group className="mb-3 d-flex flex-column">
      <Form.Label className="text-center">Avatar
      </Form.Label>
      <Image
        src={form.avatar}
        roundedCircle
        width={120}
        height={120}
        className="m-2 border border-primary border-2
                   align-self-center"
      />
      <Form.Control
        onChange={(e) => setAvatar(e.target.files[0])}
        className="w-50 align-self-center"
        type="file"
        size="sm"
      />
      <Form.Control.Feedback type="invalid">
        This file is required.
      </Form.Control.Feedback>
    </Form.Group>
    ...
  </Form>
);
```

Great! Let's add the fields for the last name and first name:

src/components/UpdateProfileForm.jsx

```
...
<Form.Group className="mb-3">
  <Form.Label>First Name</Form.Label>
  <Form.Control
    value={form.first_name}
```

```
    onChange={(e) => setForm({ ...form, first_name:
                            e.target.value })}
    required
    type="text"
    placeholder="Enter first name"
  />
  <Form.Control.Feedback type="invalid">
    This file is required.
  </Form.Control.Feedback>
</Form.Group>
<Form.Group className="mb-3">
  <Form.Label>Last name</Form.Label>
  <Form.Control
    value={form.last_name}
    onChange={(e) => setForm({ ...form, last_name:
                            e.target.value })}
    required
    type="text"
    placeholder="Enter last name"
  />
  <Form.Control.Feedback type="invalid">
    This file is required.
  </Form.Control.Feedback>
</Form.Group>
...
```

Finally, let us add the bio field and the submit button:

src/components/UpdateProfileForm.jsx

```
...
<Form.Group className="mb-3">
  <Form.Label>Bio</Form.Label>
  <Form.Control
    value={form.bio}
    onChange={(e) => setForm({ ...form, bio: e.target.value })}
    as="textarea"
```

```
    rows={3}
    placeholder="A simple bio ... (Optional)"
  />
</Form.Group>

<div className="text-content text-danger">{error &&
<p>{error}</p>}</div>

<Button variant="primary" type="submit">
  Save changes
</Button>
...
```

Great! The `UpdateProfileForm` component is written and we can use it to create the `EditProfile.jsx` page.

Creating the EditProfile page

Inside the `src/pages/` directory, create a new file called `EditProfile.jsx`. This file will contain the code for the page that will display the form to edit information about the user:

src/pages/EditProfile.jsx

```
import React from "react";
import { useParams } from "react-router-dom";
import useSWR from "swr";
import Layout from "../components/Layout";
import UpdateProfileForm from "../components/profile/
UpdateProfileForm";
import { fetcher } from "../helpers/axios";
import { Row, Col } from "react-bootstrap";

function EditProfile() {
  return (
    //JSX code
  );
}

export default EditProfile;
```

With the needed imports added, we can now add the fetching logic and the UI:

src/pages/EditProfile.jsx

```
...
function EditProfile() {
  const { profileId } = useParams();

  const profile = useSWR(`/user/${profileId}/`, fetcher);

  return (
    <Layout hasNavigationBack>
      {profile.data ? (
        <Row className="justify-content-evenly">
          <Col sm={9}>
            <UpdateProfileForm profile={profile.data} />
          </Col>
        </Row>
      ) : (
        <div>Loading...</div>
      )}
    </Layout>
  );
}
...
```

In the `EditProfile` function, we are planning to retrieve the `profileId` that will be used to fetch the up-to-date user information and pass the response to the `UpdateProfileForm` component. Naturally, we are returning the **Loading** text if the data is not pulled from the server yet. Great! Let's register this page in the `App.js` file:

src/App.jsx

```
...
<Route
  path="/profile/:profileId/edit/"
  element={
    <ProtectedRoute>
```

```
        <EditProfile />
      </ProtectedRoute>
    }
  />
  ...
```

Now, go to your profile and click on the **Edit** button. Change the information and add an avatar image to make sure everything is working.

The React application is nearly done. We have CRUD operations for authentication, posts, comments, and new users. Now, it's time to focus on the quality and maintainability of our components.

Summary

In this chapter, we added CRUD operations for the user in the React application. We explored how powerful and simple it is to handle media uploading in Django and how to create a form that can accept file uploads to a remote server. We have also added new components to the React application for better navigation and exploration of other profiles. We are done implementing most features of our application.

In the next chapter, we will learn how to write tests for a React frontend application.

Questions

1. What is a `formData` object?
2. What is the `MEDIA_URL` setting usage in Django?
3. What is the `MEDIA_ROOT` setting usage in Django?

11

Effective UI Testing for React Components

We have already been introduced to testing with Python and Django in *Chapter 5, Testing the REST API*. In this chapter, the context is different as we will work with JavaScript and React to test the frontend components we have designed and implemented. This chapter will show you what to test in a frontend application and how to write tests for React UI components.

In this chapter, we will cover the following topics:

- Component testing in React
- Jest and the **React Testing Library** (**RTL**)
- Testing form components
- Testing `post` components
- Snapshot testing

Technical requirements

Make sure to have VS Code and an updated browser installed and configured on your machine. You can find the code from this chapter at `https://github.com/PacktPublishing/Full-stack-Django-and-React/tree/chap11`.

Component testing in React

We already understand that the frontend is the client-side section of an application. Concerning the tests we wrote in *Chapter 5, Testing the REST API*, in our Django applications, we mostly tested whether the database stored the right data passed to the viewsets, serializers, and models. However, we didn't test the user interface.

As a React developer, you might be thinking: what do I test in my frontend application? Well, let's respond to this question by understanding why a frontend test is needed and what needs to be tested.

The necessity of testing your frontend

When developing an application, it's important to ensure that your application works as expected in a production environment.

The frontend also represents the interface the user will use to interact with your backend. For a good user experience, it's crucial to write tests that ensure that your components are behaving as expected.

What to test in your React application

If you are coming from a backend viewpoint, you might be a little bit confused about what to test in your frontend application. From a basic aspect, it's not different from testing your backend. If you have classes or methods in your application, you can write tests. Frontend testing includes testing different aspects of the UI such as formatting, visible text, graphics, and the functional parts of the applications such as buttons, forms, or clickable links.

Now, the difference is that your React frontend is made of UI components, taking props for displaying data to the user. The React ecosystem provides testing tools that easily help you write tests for your components.

In the next section, we will start with a small introduction to Jest and the RTL and then we will write tests for our authentication forms.

Jest, the RTL, and fixtures

Jest is a JavaScript framework for writing, running, and structuring tests. It comes with all the tools needed to check code coverage, easily mock functions, and imported functions, and write simple and great exceptions. The RTL is a library for actually testing React applications. It focuses on testing components from a user experience point of view rather than testing the implementation and logic of the React components themselves.

> **Important note**
>
> When writing tests, you will often need to ensure that some values or variables meet certain conditions. This was done in *Chapter 5, Testing the REST API*, of this book, using `assert` when writing tests for the Django application using pytest. Working with Jest, the term changes from assertion to exceptions. When doing frontend testing with Jest, we are expecting the value to meet a condition. For example, if the user enters and clicks on a button that will reset a form, we expect the form to be reset after the click action is made on the button.

The RTL is not separated from Jest as you need both to write tests for your frontend application. Jest will help you write the testing blocks while the RTL will provide tools to select components, render the components, and trigger common user events such as clicking and typing. These tools are already installed by default when creating a React project, so there is no need to add other packages.

The only packages we will need are `faker.js` and the JavaScript `uuid` package to generate UUID4 identifiers. Faker is a JavaScript package used to generate fake, but realistic, data. In the React project, use the following command to install the package as a development dependency:

```
yarn add @faker-js/faker uuid -dev
```

With the packages installed, we can now add some important fixtures for the components we are going to test in the next lines.

Writing testing fixtures

In the `src/helpers` directory, create a new directory called `fixtures`. This directory will contain JavaScript files containing functions that return fixtures that can be used for testing.

We'll start by writing fixtures for a user. So, in the `fixtures` directory, create a new file called `user.js`. This file will contain code for a function that returns realistic data for a user object. Let's start with the imports of functions from the `faker.js` and `uuid` packages to create a fixture:

src/helpers/fixtures/user.js

```
import { faker } from "@faker-js/faker";
import { v4 as uuid4 } from "uuid";

function userFixtures() {
...
}
export default userFixtures;
```

With the imports and the structure of the `userFixtures` function written, we can now return the object fixture:

src/helpers/fixtures/user.js

```
...
function userFixtures() {
 const firstName = faker.name.firstName();
```

```
  const lastName = faker.name.lastName();

  return {
    id: uuid4(),
    first_name: firstName,
    last_name: lastName,
    name: firstName + " " + lastName,
    post_count: Math.floor(Math.random() * 10),
    email: `${firstName}@yopmail.com`,
    bio: faker.lorem.sentence(20),
    username: firstName + lastName,
    avatar: null,
    created: faker.date.recent(),
    updated: faker.date.recent(),
  };
}
...
```

Faker provides a lot of modules with methods to return data. In the previous code block, we are working with `faker.name` to generate random names, `faker.lorem` to generate random lorem texts, and `faker.date` to generate a recent date. The object returned by `userFixtures` now has the closest structure to a user object returned by the Django API we have created, and this is exactly what we want.

Before diving into component testing, let's make sure our testing environment is well configured.

Running the first test

When a React application is created, the `App.js` file comes with a test file called `App.test.js`, which you can see here:

src/App.test.js

```
import { render, screen } from "@testing-library/react";
import App from "./App";

test("renders learn react link", () => {
  render(<App />);
  const linkElement = screen.getByText(/learn react/i);
```

```
  expect(linkElement).toBeInTheDocument();
});
```

Let me explain the code. Here, we are importing the `render` and `screen` methods from the RTL. These modules will be used to render a component and make interactions with the components easier by providing methods to select DOM elements, respectively.

Next, we have the `test` method. It's simply a Jest keyword used to write tests. It takes two parameters: a string describing the test, and the callback function where you write the testing logic. Inside the callback function, the `App` component is rendered first. Then, `linkElement` is retrieved from the screen by using the `learn react` text. Once it's retrieved, we can then check whether the `linkElement` exists in the rendered document.

Let's run this test with the following command:

```
yarn test
```

You should have a similar output in the terminal.

```
Test Suites: 1 failed, 1 total
Tests:       1 failed, 1 total
Snapshots:   0 total
Time:        2.6 s
Ran all test suites.

Watch Usage: Press w to show more.
```

Figure 11.1 – Running the yarn test command

The test has failed. But why? You can somewhat see why in the preceding output.

```
 FAIL  src/App.test.js
  × renders learn react link (134 ms)

  ● renders learn react link

    useRoutes() may be used only in the context of a <Router> component.
```

Figure 11.2 – Reason for failing the App.js test

The `App` component in our project uses `react-router-dom` components, such as `Routes`, that in turn use the `useRoutes` Hook. This Hook makes use of the context that a router component provides, so we need to wrap it inside a Router, in this case, the `BrowserRouter` component. Let's correct this, but let's also change the text from which we will retrieve the link element:

src/App.test.js

```
import { render, screen } from "@testing-library/react";
import App from "./App";
import { BrowserRouter } from "react-router-dom";

test("renders Welcome to Postagram text", () => {
  render(
    <BrowserRouter>
      <App />
    </BrowserRouter>
  );
  const textElement =
    screen.getByText(/Welcome to Postagram!/i);
  expect(textElement).toBeInTheDocument();
});
```

Now, run the tests again and everything should work correctly:

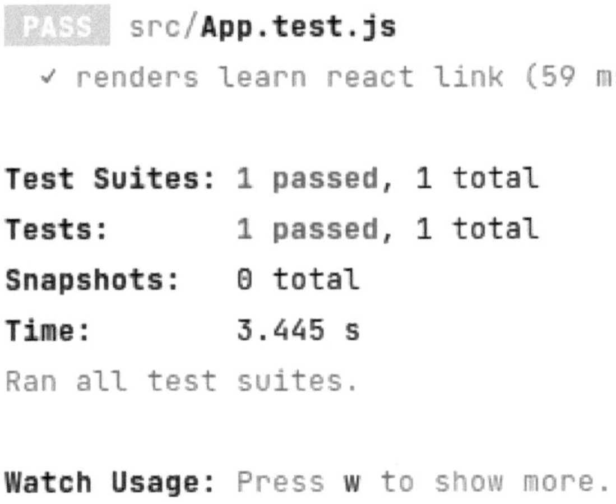

Figure 11.3 – Passing tests

But we still have a problem. A lot of components in the React application use Hooks from the `react-router-dom` library. That means that for each test, we will need to wrap the component inside `BrowserRouter`. Following the DRY principle, let's rewrite the render method from the RTL to automatically wrap our components inside the `BrowserRouter`.

Extending the RTL render method

Inside the `src/helpers` directory, create a file called `test-utils.jsx`. Once the file is created, add the following lines of code:

src/helpers/test-utils.jsx

```
import React from "react";
import { render as rtlRender } from "@testing-library/react";
import { BrowserRouter } from "react-router-dom";

function render(ui, { ...renderOptions } = {}) {
  const Wrapper = ({ children }) =>
    <BrowserRouter>{children}</BrowserRouter>;
  return rtlRender(ui, { wrapper: Wrapper, ...renderOptions
    });
}

export * from "@testing-library/react";
export { render };
```

In the code, we first import the needed tools. Notice the import of the render method as `rtlRender`? It's for the purpose of avoiding a naming collision as we are also writing a render function. Next, we create a function called `Wrapper`, where we pass the children's argument, then wrap it inside a `BrowserRouter` component. Then, we return a render object with the UI, `wrapper`, and other render options if specified.

> **Important note**
>
> The render method from the RTL provides render options other than the wrapper. You can also pass a container, queries, and a lot more. You can check more rendering options in the official documentation at `https://testing-library.com/docs/react-testing-library/api/#render-options`.

Now, let's use this method in `App.test.js`:

src/App.test.js

```
import { render, screen } from "@testing-library/react";
import App from "./App";

test("renders Welcome to Postagram text", () => {
 render(<App />);
...
});
```

Run the testing command again and everything should be green. With the testing environment ready and set up to quickly write tests, we can now move on to testing the components of the React project.

> **Important note**
>
> While running tests, you might encounter an error coming from the `axios` package. At the time of writing of this book, we are using the 0.26.0 version of `axios` to avoid bugs when running tests. You can also modify the test command in the `package.json` file to the following: `"test": "react-scripts test --transformIgnorePatterns "node_modules/(?!axios)/""`. Learn more about this issue at `https://github.com/axios/axios/issues/5101`.

Testing authentication components

Testing forms in React might seem complicated but it is quite simple when using Jest and the RTL. We will begin writing tests in the React project, starting with the authentication components. I'll show you how you can write a test for the **Login** form, and after that, you should be able to write the registration form test suite.

For a better structured code base, create a new directory called `__tests__` in the `src/components/authentication` directory. This directory will contain tests for the components in the `components/authentication` directory. Inside the newly created directory, create a file called `LoginForm.test.js` and add the following code:

src/components/authentication/__tests__/LoginForm.test.js

```
import { render, screen } from "../../../helpers/test-utils";
import userEvent from "@testing-library/user-event";
import LoginForm from "../LoginForm";
```

```
import { faker } from "@faker-js/faker";
import userFixtures from "../../../helpers/fixtures/user";

const userData = userFixtures();

test("renders Login form", async () => {
...
});
```

In the preceding code, we have added the required imports to write the test and defined the structure of the testing function. We will first render the `LoginForm` component and set up the user object to trigger user behavior events using `userEvent` method: `src/components/authentication/__tests__/LoginForm.test.js`:

```
...
test("renders Login form", async () => {
  const user = userEvent.setup();
  render(<LoginForm />);
...
```

Important note

`userEvent` and `fireEvent` are both methods used for simulating user interactions with a website in the context of testing. They can be used to test the behavior of a website when a user performs certain actions, such as clicking a button or filling out a form.

`userEvent` is a method provided by the `@testing-library/user-event` library, which is designed to make it easier to test user interactions with a website. It is a utility function that simulates user events by using the `fireEvent` method provided by the `@testing-library/react` library. `userEvent` allows you to specify the type of event you want to simulate, such as a click or a keypress, and it will automatically dispatch the appropriate event for you.

`fireEvent` is a method provided by the `@testing-library/react` library, which can be used to dispatch events to a DOM element. It allows you to specify the type of event you want to dispatch, as well as any additional event data that you want to include. `fireEvent` is a more low-level method than `userEvent`, and it requires you to manually specify the details of the event you want to dispatch.

After that, we can start by testing that the form and the inputs are rendered in the document:

src/components/authentication/__tests__/LoginForm.test.js

```
test("renders Login form", async () => {
...
 const loginForm = screen.getByTestId("login-form");
 expect(loginForm).toBeInTheDocument();

const usernameField = screen.getByTestId("username-field");
expect(usernameField).toBeInTheDocument();

const passwordField = screen.getByTestId("password-field");
expect(passwordField).toBeInTheDocument();
...
```

Then, we can ensure that the inputs can receive texts and values as we have already selected the username and password fields:

src/components/authentication/__tests__/LoginForm.test.js

```
test("renders Login form", async () => {
...
 const password = faker.lorem.slug(2);
 await user.type(usernameField, userData.username);
 await user.type(passwordField, password);

 expect(usernameField.value).toBe(userData.username);
 expect(passwordField.value).toBe(password);
});
```

If you run the test command again, it will fail. That's normal, as here we are retrieving elements using the `getByTestId` method. The RTL looks in the rendered DOM for an element with a `data-testid` attribute with the value passed to the `screen.getByTestId` function. We need to add the attribute to the elements we want to select and test.

To do so, in `src/components/authentication/LoginForm.js`, add the following `data-testid` attributes:

src/components/authentication/LoginForm.js

```
function LoginForm() {
...
 return (
   <Form
     id="registration-form"
     className="border p-4 rounded"
     noValidate
     validated={validated}
     onSubmit={handleSubmit}
     data-testid="login-form"
   >
...
       <Form.Label>Username</Form.Label>
       <Form.Control
         value={form.username}
         data-testid="username-field"
...
     <Form.Group className="mb-3">
       <Form.Label>Password</Form.Label>
       <Form.Control
         value={form.password}
         data-testid="password-field"
...
```

Once done, re-run the testing command. Everything should work.

The next step is to write tests for the registration form component. It'll be similar to the tests on the login form component, so you can handle this small exercise. You can find the solution at `https://github.com/PacktPublishing/Full-stack-Django-and-React/blob/main/social-media-react/src/components/authentication/__tests__/RegistrationForm.test.js`.

> **Important note**
>
> JavaScript also possesses default naming conventions for the testing files. The naming conventions for test files in a JavaScript project are as follows:
>
> - `<TestFileName>.test.js`
>
> - `<TestFileName>.spec.js`

Now that we have had a solid introduction to testing in React, let's continue with writing tests for the Post components.

Testing Post components

The functionalities to create, read, update, and delete posts are core features of the Postagram application, so it's important to make sure that they work as expected. Let's start with a simple test for the `Post` component.

Mocking the localStorage object

Before writing a test for the `Post` component, it's important to understand how the Post components work. Basically, it takes a prop called `post` and makes a call to `localStorage` to retrieve information about the user. Unfortunately, `localStorage` can't be mocked by Jest. There are a lot of workarounds to allow your tests to work with `localStorage` and to make it simple and have less boilerplate, we'll use the `jest-localstorage-mock` JavaScript package. The package can be used with Jest to run frontend tests that rely on `localStorage`. To add the package, add the following line to the file:

```
yarn add --dev jest-localstorage-mock
```

Once the package is installed, we need to do some configurations. In the `src/setupTests.js` file, add this line to load the `jest-localstorage-mock` package:

src/setupTests.js

```
...
require('jest-localstorage-mock');
```

After that, override the default Jest configuration in the `package.json` file:

package.json

```
...
{
  "jest": {
```

```
    "resetMocks": false
  }
}
...
```

With the configuration ready, we can move to add a function to generate post fixtures.

Writing post fixtures

In the `src/helpers/fixtures` directory, create a new file called `post.js`. This file will contain a function that returns fake data from a post object.

We will start writing the code in this file by adding the imports and defining the `postFixtures` function that will return a generated post object:

src/helpers/fixtures/post.js

```
import { faker } from "@faker-js/faker";
import { v4 as uuid4 } from "uuid";
import userFixtures from "./user";


function postFixtures(isLiked = true, isEdited = false, user =
undefined) {
...
}

export default postFixtures;
```

Let's add the body of the `postFixtures` function:

src/helpers/fixtures/post.js

```
...
function postFixtures(isLiked = true, isEdited = false, user =
undefined) {
 return {
   id: uuid4(),
   author: user || userFixtures(),
   body: faker.lorem.sentence(20),
```

```
    edited: isEdited,
    liked: isLiked,
    likes_count: Math.floor(Math.random() * 10),
    comments_count: Math.floor(Math.random() * 10),
    created: faker.date.recent(),
    updated: faker.date.recent(),
  };
}
```

Here, we are passing either a generated `userFixtures` or a user object if it's defined. This is important if we want to make sure that the author of the post is the same user registered in `localStorage`.

With the post fixtures written, we can write the test suite for the `Post` component.

Writing tests for the Post component

To write the test suite in the `src/components/posts` directory, create a new folder called `__tests__`. Inside the newly created folder, add a new file called `Post.test.js`. Inside, add the imports, create the data we need, and set user data returned by the `userFixtures` function in the local storage using the `setUserData` function:

src/components/posts/__tests__/Post.test.js

```
import { render, screen } from "../../../helpers/test-utils";
import Post from "../Post";
import { setUserData } from "../../../hooks/user.actions";
import userFixtures from "../../../helpers/fixtures/user";
import postFixtures from "../../../helpers/fixtures/post";

const userData = userFixtures();

const postData = postFixtures(true, false, userData);

beforeEach(() => {
 // to fully reset the state between __tests__, clear the
 // storage
 localStorage.clear();
 // and reset all mocks
```

```
  jest.clearAllMocks();

  setUserData({
    user: userData,
    access: null,
    refresh: null,
  });
});
```

The `beforeEach` method is a Jest method that runs before every test. It takes a callback function as a parameter, where you can execute lines of code that should run before the tests. Here, we are clearing the local storage first to avoid memory leaking (with `localStorage.clear`) and finally, we set user data retrieved from the `userFixtures` function in the local storage.

> **Important note**
>
> A memory leak occurs when a program creates a memory in heap and forgets to delete it. In a worst-case scenario, if too much memory is allocated and not used correctly, this can reduce the computer's performance.

Let's write the test for the `Post` component now:

src/components/posts/__tests__/Post.test.js

```
...
test("render Post component", () => {
  render(<Post post={postData} />);
  const postElement = screen.getByTestId("post-test");
  expect(postElement).toBeInTheDocument();
});
```

If you run the test command, it'll fail. This is normal because there is no `data-testid` attribute with the value `post-test` set in the JSX of the `Post` component. Let's fix this by adding a `data-testid` attribute in the `Post` component :

src/components/posts/Post.jsx

```
...
function Post(props) {
```

```
...

  return (
    <>
      <Card className="rounded-3 my-4"
        data-testid="post-test">
...
    </>
  );
}

export default Post;
```

Run the testing command again and everything should be green. Let's move on to actually writing a test for the `CreatePost` component.

Testing the CreatePost component

In the `src/components/posts/__tests__` directory, create a new file called `CreatePost.test.js`. We'll start with the necessary imports and the definition of the `test` function:

src/components/posts/__tests__/CreatePost.test.js

```
import { render, screen, fireEvent } from "../../../helpers/
test-utils";
import userEvent from "@testing-library/user-event";
import CreatePost from "../CreatePost";
import { faker } from "@faker-js/faker";


test("Renders CreatePost component", async () => {
...
});
```

You can notice the introduction of the `async` keyword before the callback function. To create a post, the user performs typing operations on text inputs and finally a button click to submit the post. These actions are asynchronous. The functions, such as `fireEvent`, that we will use to simulate user interactions should be used in an asynchronous scope.

Before writing the test logic, let's remember how the `CreatePost` component works:

1. The user clicks on the input to add a new post.
2. A modal is shown containing a form where the user can enter the text of the post. Meanwhile, the submit button is disabled.
3. Once there is enough text in the field, the submit button is enabled and the user can click to send the post.

We must ensure that we respect this logic when writing the tests. Now, let's start writing the tests.

First, we render the form that displays the form modal to create a post:

src/components/posts/__tests__/CreatePost.test.js

```
test("Renders CreatePost component", async () => {
  const user = userEvent.setup();
  render(<CreatePost />);


  const showModalForm =
    screen.getByTestId("show-modal-form");
  expect(showModalForm).toBeInTheDocument();
});
```

We can now simulate a click event using `fireEvent.click` on `showModalForm` to display the form for creating a post:

src/components/posts/__tests__/CreatePost.test.js

```
...
  // Clicking to show the modal
  fireEvent.click(showModalForm);

  const createFormElement =
    screen.getByTestId("create-post-form");
  expect(createFormElement).toBeInTheDocument();
...
```

We then make sure that the body field is rendered and the submit button is disabled:

src/components/posts/__tests__/CreatePost.test.js

```
...
  const postBodyField =
    screen.getByTestId("post-body-field");
  expect(postBodyField).toBeInTheDocument();

  const submitButton =
    screen.getByTestId("create-post-submit");
  expect(submitButton).toBeInTheDocument();

  expect(submitButton.disabled).toBeTruthy();
  ...
```

After that, we can then type some text in the body field, test whether the text typed is what we expect, and ensure that the button is enabled after that:

src/components/posts/__tests__/CreatePost.test.js

```
  ...
  const postBody = faker.lorem.sentence(10);
  await user.type(postBodyField, postBody);

  // Checking if field has the text and button is not
  // disabled

  expect(postBodyField.value).toBe(postBody);
  expect(submitButton.disabled).toBeFalsy();
});
```

Great! We have a solid testing suite and we can now add the `data-testid` attributes to the `CreatePost` component to make the tests pass:

src/components/posts/CreatePost.jsx

```
function CreatePost() {
...
```

```
  return (
    <>
      <Form.Group className="my-3 w-75">
        <Form.Control
          className="py-2 rounded-pill border-primary
                     text-primary"
          data-testid="show-modal-form"
...
        <Modal.Body className="border-0">
          <Form
            noValidate
            validated={validated}
            onSubmit={handleSubmit}
            data-testid="create-post-form"
          >
            <Form.Group className="mb-3">
              <Form.Control
                name="body"
                data-testid="post-body-field"
...
        </Modal.Body>
        <Modal.Footer>
          <Button
            variant="primary"
            onClick={handleSubmit}
            disabled={!form.body}
            data-testid="create-post-submit"
...
    </>
  );
}
```

Run the test command again and everything should work. The next step is to write unit tests for the `UpdatePost` component.

Testing the UpdatePost component

In the `src/components/posts/__tests__` directory, create a new file called `UpdatePost.test.js`. Let's start with the necessary imports and the definition of the `test` function:

src/components/posts/__tests__/UpdatePost.test.js

```
import { render, screen, fireEvent } from "../../../helpers/
test-utils";
import userEvent from "@testing-library/user-event";
import UpdatePost from "../UpdatePost";
import userFixtures from "../../../helpers/fixtures/user";
import postFixtures from "../../../helpers/fixtures/post";
import { faker } from "@faker-js/faker";

const userData = userFixtures();
const postData = postFixtures(true, false, userData);

test("Render UpdatePost component", async () => {
...
});
```

Before writing the test logic, let's remember how the `UpdatePost` component works from a user's perspective:

1. The user clicks on the drop-down item to modify a post.
2. A modal is shown containing a form where the user can modify the text of the post.
3. After the modification, the user can submit the form with the updated post.

We must ensure that we respect that logic when writing the tests.

So, first, we render the form that displays the form modal to update a post:

src/components/posts/__tests__/UpdatePost.test.js

```
test("Render UpdatePost component", async () => {
  const user = userEvent.setup();
  render(<UpdatePost post={postData} />);
```

```
  const showModalForm =
    screen.getByTestId("show-modal-form");
  expect(showModalForm).toBeInTheDocument();
...
```

We then want to trigger a click event to display the modal with the form to update the post:

src/components/posts/__tests__/UpdatePost.test.js

```
...
  fireEvent.click(showModalForm);

  const updateFormElement =
    screen.getByTestId("update-post-form");
  expect(updateFormElement).toBeInTheDocument();
...
```

We then select the post body field and the submit button to ensure that they are rendered:

src/components/posts/__tests__/UpdatePost.test.js

```
...
  const postBodyField =
    screen.getByTestId("post-body-field");
  expect(postBodyField).toBeInTheDocument();

  const submitButton =
    screen.getByTestId("update-post-submit");
  expect(submitButton).toBeInTheDocument();
...
```

After that, we can now trigger a typing event in the post body field and ensure that the user is submitting the right data:

src/components/posts/__tests__/UpdatePost.test.js

```
...
  const postBody = faker.lorem.sentence(10);
```

```
  await user.type(postBodyField, postBody);


  // Checking if field has the text and button is not
  // disabled

  expect(postBodyField.value).toBe(postData.body +
    postBody);
  expect(submitButton.disabled).toBeFalsy();
});
```

The next step is now to add the `data-testid` attributes on the post form, the post body input, and the submit button in the `UpdatePost` component to make the tests pass:

src/components/posts/UpdatePost.jsx

```
...
function UpdatePost(props) {
...
  return (
    <>
      <Dropdown.Item data-testid="show-modal-form"
        onClick={handleShow}>
...
        <Modal.Body className="border-0">
          <Form
            noValidate
            validated={validated}
            onSubmit={handleSubmit}
    data-testid="update-post-form"
    >

            <Form.Group className="mb-3">
              <Form.Control
                name="body"
                value={form.body}
                data-testid="post-body-field"
```

```
...
        </Modal.Body>
        <Modal.Footer>
          <Button
            data-testid="update-post-submit"
...
```

Run the test command again and everything should work.

With this introduction to complex tests with Jest and the RTL, you can easily write the tests for the comment's components. You can find the solution for these tests at `https://github.com/PacktPublishing/Full-stack-Django-and-React/tree/main/social-media-react/src/components/comments/__tests__`. Good luck!

In the next section, we will discover what snapshot testing is.

Snapshot testing

Snapshot tests are a very useful tool when you want to make sure that your UI does not change unexpectedly. A snapshot test case follows these steps:

- It renders the UI component.
- It then takes a snapshot and compares it to a reference snapshot file stored alongside the test file.
- If both states are the same, the snapshot test is successful. Otherwise, you will get errors and need to decide whether you need to update the snapshot tests or fix your components.

Snapshot tests are great to prevent UI regression and ensure that the application adheres to the code quality and values of your development team.

There is a minor setback with snapshot tests, however. Snapshot testing doesn't work best with dynamic components. For example, the `Post` component uses `timeago` to display a human-readable time. This means that a snapshot of this component at time `t` will be different from a snapshot of the same component at time `t + 1`. However, there are some static components in the React application such as `LoginForm`, `RegistrationForm`, `ProfileDetails`, `ProfileCard`, `CreatePost`, and much more.

For the sake of simplicity, we will write a snapshot test for the `ProfileCard` components, which are straightforward and can be replicated easily.

In the `src/components/profile` directory, create a new directory called `__tests__`. Then, create a new file called `ProfileCard.test.js`. For a snapshot test, we don't want the data to change so we will use a static user fixture because using `userFixtures` to generate a fixture will create random data every time a snapshot test is run. In the newly created file, let's add the imports needed to create a snapshot test and define a fixture object called `userData`:

src/components/profile/__tests__/ProfileCard.test.js

```
import { render, screen } from "../../../helpers/test-utils";
import TestRenderer from "react-test-renderer";
import ProfileCard from "../ProfileCard";
import { BrowserRouter } from "react-router-dom";

const userData = {
 id: "0590cd67-eacd-4299-8413-605bd547ea17",
 first_name: "Mossie",
 last_name: "Murphy",
 name: "Mossie Murphy",
 post_count: 3,
 email: "Mossie@yopmail.com",
 bio: "Omnis necessitatibus facere vel in est provident
       sunt tempora earum accusantium debitis vel est
       architecto minima quis sint et asperiores.",
 username: "MossieMurphy",
 avatar: null,
 created: "2022-08-19T17:31:03.310Z",
 updated: "2022-08-20T07:38:47.631Z",
};
```

With the needed imports added and the `userData` fixtures written, we can now write the testing function:

src/components/profile/__tests__/ProfileCard.test.js

```
...
test("Profile Card snapshot", () => {
 const profileCardDomTree = TestRenderer.create(
   <BrowserRouter>
     <ProfileCard user={userData} />
   </BrowserRouter>
 ).toJSON();
```

```
  expect(profileCardDomTree).toMatchSnapshot();
});
```

If you run the test command, you'll notice that a snapshot directory is created in the `__tests__` directory:

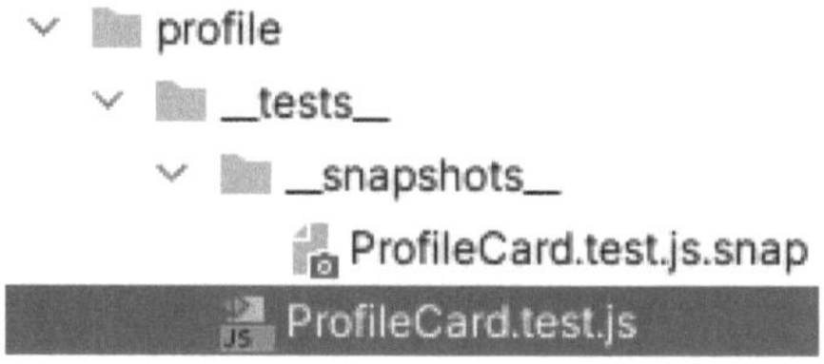

Figure 11.4 – Snapshots directory created

If you check the content of `ProfileCard.test.js.snap`, it is basically the rendered code of the `ProfileCard` component. The content of this file will be compared each time the test function for the snapshot test runs.

Now have covered the essential unit tests for a React application, we are mostly done adding features to the application. Our full stack application is now ready for production! Yay, but don't celebrate too soon. We still need to prepare our application for production in terms of the security, quality, and performance aspects and this is what we'll be doing in *Part 3* of this book.

Summary

In this chapter, you have learned about frontend unit testing. We discovered why it is important to write unit tests in the frontend application and what to test exactly. We have also written tests for components in the Postagram application, seen how we can extend testing tools modules and methods, how to write generate fixtures for the tests, and how to make the tests closer to user interactions by triggering user events. We have also made some introductions to snapshot testing.

The next chapters in *Part 3* of this book will focus on deploying the backend and the frontend on the cloud using AWS services, GitHub, and GitHub Actions. Lastly, we will see how to improve the full stack application in terms of performance, security, and quality.

Questions

1. What is the render method of the RTL?
2. What is Jest?
3. What is the role of the `data-tested` attribute?
4. What are the drawbacks of snapshot testing?
5. What are the modules used to trigger user events on a React test suite?

Part 3: Deploying Django and React on AWS

Deployment is one of the last important steps in software development. Your application is running locally and everything is working fine. But how do you get your code on a public server? How do you host your frontend? How do you make changes to your code and make deployment and testing automatic? In this part of the book, we'll explore topics such as CI/CD, GitHub, Docker, and the best deployment practices while deploying the Django application on AWS EC2 and the React application on AWS S3. We'll also talk about security and performance.

This section comprises the following chapters:

12

Deployment Basics – Git, GitHub, and AWS

It's nice to develop an application with a functioning backend and a nice, flexible frontend on your machine. Still, if you want your application to be used publicly, you need to deploy the application to production. From this chapter to the last one, you will learn how to prepare the application we've built for deployment, deploy the backend on **Amazon Web Services** (**AWS**) and the frontend on Vercel, and finally, go through some security and performance optimizations.

In this chapter, we will learn deployment basics such as jargon and concepts to understand before going further. We will be learning about the following topics:

- Basics of software deployment
- Tools and methods of web application deployment
- Platforms for web application deployment

Technical requirements

For this chapter, you will need to have Git installed on your machine. If you are on Linux or macOS, it will come by default. You can check its existence with the following command in the terminal:

```
git -version
```

Otherwise, feel free to download the right version at `https://git-scm.com/downloads`.

After the installation, let's configure Git if not done yet. In a terminal, enter the following configuration commands to set the username (usually the username on your GitHub account) and the email address (usually the email address on your GitHub account):

```
git config --global user.name "username"
git config --global user.email "email@address.com"
```

You will also need an active GitHub account. You can register on the official website at `https://github.com/`. As we will also be deploying the application on a remote AWS server, you will need an AWS account that can be created at `https://portal.aws.amazon.com/billing/signup`. If you don't have an AWS account, you can still use any **virtual private server** (**VPS**) or **virtual private cloud** (**VPC**) you have online. However, this chapter will also document how to create a VPC instance using AWS and how to upload the code and serve the Django API.

Basics of software deployment

Software deployment concerns all the activities that make a software system available to consumers. The term *software deployment* is also commonly described as application deployment. Following the best software deployment practices will ensure that all applications deployed operate smoothly and work as expected.

There are several benefits of software deployment, such as:

- **Saved time**: A good software deployment process can be configured to only take a few minutes. This saves time for compiling and distribution to the users.
- **Increased security**: Deploying your application in a structured manner rather than doing it manually or for individual users means you ensure the security of the application and not only the security of the application on every user's device.
- **Better monitoring**: Deploying an application on production servers helps provide more control and data on what is working from the user's end.

With software deployment defined, we will dive deeper into the tools and methods used for web application deployment.

Tools and methods of web application deployment

Deploying a web application for production has drastically evolved over the years. From manual deployment to automated deployment techniques, web application deployment has advanced, making the process more secure, smooth, and as fast as possible. There are many tools for web application deployment, but in this book, we will focus on the automated tools and configure the Django project and the React project for automated deployments when pushes are made on the remote repository of the code.

But where will the code be pushed first? Let's start describing and learning how to use the tools for our full stack application deployment, starting with Git and GitHub.

Using Git and GitHub

Git is a popular tool used for source code version control and collaboration. It not only helps the user keep track of changes made to the code but also allows developers to work through small or large code bases, with collaboration made easier. In the following subsections, we will initialize a Git repository in the backend project, commit the changes, and then push the changes to a remote repository on GitHub.

Creating a Git repository

Open a new terminal in the directory where you created the Django project and enter the following command:

```
git init
```

This command will create an empty `.git/` directory in the current directory: this is a Git repository. This repository tracks all changes made to files in the project, helping build a history of changes made, with details on the files changed, the name of the person making the changes, and much more information.

After the initialization, we will need to ignore some files in the project. We are talking about files such as `.pycache`, `.env`, and the virtual environment directories. After all, we don't want important information such as secret environment variables to be available in the project or useless cache files to be present in the changes.

Inside the directory of the Django API, create a new file called `.gitignore`. This file tells Git which files and directories to ignore when tracking changes:

.gitignore

```
__pycache__
venv
env
.env
```

These files and directories in the preceding code will be ignored. Next, we will add the change in the directory to the staging area. The staging area allows you to group related changes before committing them to the project history. As we have successfully added a `.gitignore` file, we can freely run the `git add` command:

```
git add .
```

The dot (`.`) at the end of the command tells Git to only look for changed files in the current directory. To have a look at the changes to be committed to the Git history, run the following command:

```
git status
```

The `git status` command is used to show the state of the working directory and also the staging area. Using the command, you can see changes that are tracked or not. The following figure shows an example of the output you should have:

```
(venv) koladev@koladev123xxx:~/Downloads/Full-stack-Django-and-React-main$ git status
On branch master

No commits yet

Changes to be committed:
  (use "git rm --cached <file>..." to unstage)
        new file:   .gitignore
        new file:   CoreRoot/__init__.py
        new file:   CoreRoot/asgi.py
        new file:   CoreRoot/settings.py
        new file:   CoreRoot/urls.py
        new file:   CoreRoot/wsgi.py
        new file:   LICENSE
        new file:   README.md
        new file:   conftest.py
        new file:   core/__init__.py
        new file:   core/abstract/__init__.py
```

Figure 12.1 – Running the git status command

We can now run the `git commit` command. A commit is an operation that writes the latest changes of the source code to the version control system history. In our case, with `git commit` command will save the changes to the local repository:

```
git commit
```

The preceding command will prompt you to a text editor in the terminal or an app, depending on your system. Either way, you will need to enter a message. It's important to enter a meaningful message because this message will be shown in the history of changes made to the source code. You can enter the following line if you want:

```
Initialize git in API project
```

After saving the message, you can check the Git history with the `git log` command:

```
git log
```

You will have something similar to the following figure:

```
(venv) koladev@koladev123xxx:~/Downloads/Full-stack-Django-and-React-main$ git log
commit 7fab9bf8a03f6e2bd42291214168f9c7c3d7c308 (HEAD -> master)
Author: koladev <onaelmangabo@gmail.com>
Date:   Sun Sep 18 01:09:45 2022 +0100

    Initialize git in API project
```

Figure 12.2 – Writing a commit message

> **Important note**
>
> Writing meaningful commit messages is important, particularly in a team or a collaborative environment. You can read more about commit messages at `https://www.conventionalcommits.org/en/v1.0.0/`.

The project repository has been initialized locally; however, we want the code on GitHub. The next section will show you how to upload your code on GitHub.

Uploading code on GitHub

GitHub is a code hosting platform for collaboration and version control. It helps developers around the world work together on projects and is actually the code hosting platform for the majority of popular open source projects.

On your GitHub account dashboard, on the navigation bar, create a new repository:

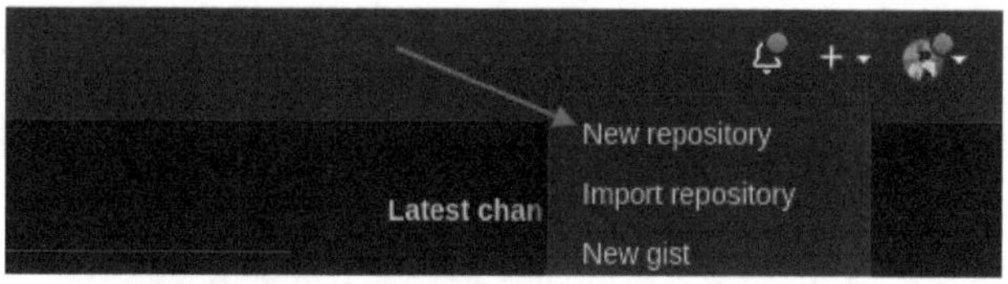

Figure 12.3 – Creating a repository on GitHub

Once it's done, you will be redirected to a new page to enter basic information about the repository, such as the name of the repository and a description, stating if the repository is public or private, and adding a license or a `.gitignore` file. The repository name is required, and the other pieces of information are optional.

You can now create the repository, and you will have a similar page to this:

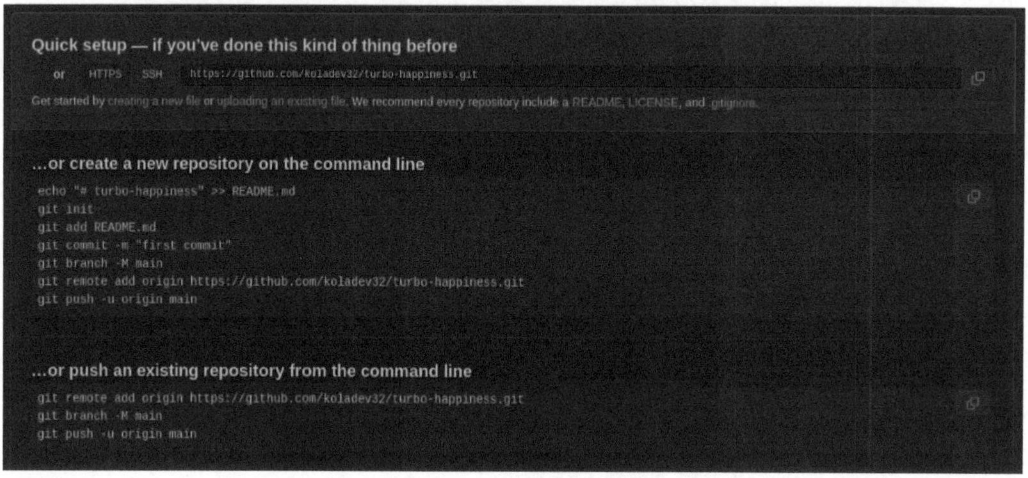

Figure 12.4 – Repository created

We have an existing repository, and we want to push it to the GitHub platform. Let's follow the steps for **...or push an existing repository from the command line**. Inside the directory of your backend project, open a new terminal, and let's enter the shell commands:

```
git remote add origin your_repository_git_url
```

The `git remote` command allows you to create, view, and delete connections to Git repositories hosted on the internet or another network. In the preceding command, we are adding a remote repository URL of the GitHub repository. Let's change the name of the branch we are working on:

```
git branch -M main
```

By default, when a repository is created using Git on a local machine, the branch of work is called master. What is a branch in Git?

Well, it is just a separate version of the main repository. This allows multiple developers to work on the same project. For example, if you are working with a backend developer who wants to add support for file uploading on posts and comments, instead of working directly on the main branch, the developer can create a new branch (`feature/images-post`) from the main branch. After the work is done on this branch, the `feature/images-post` branch can be merged with the main branch.

With the main branch created, we can now push the changes to GitHub:

```
git push -u origin main
```

The `git push` command is used to upload local repository changes on the source code to a remote repository. In your case, the command will push the current code to your GitHub repository URL.

Reload the repository page on GitHub, and you will see something similar to this:

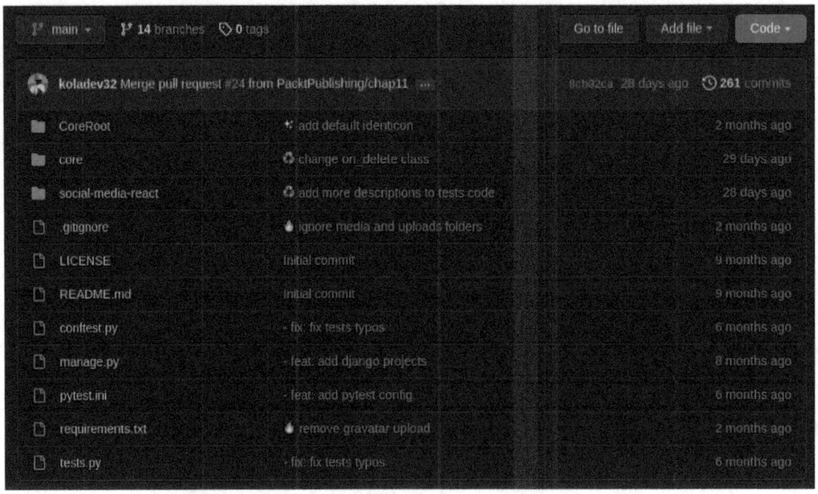

Figure 12.5 – Code pushed to the repository

And voilà! We have the code uploaded on GitHub. But this is just the code. What if you can have this running on a remote server that you can access from anywhere?

Let's talk about platforms for web application deployment and deploy the Django backend on AWS.

Platforms for web application deployment

With the complexity of software development increasing and more innovative and data-intensive applications evolving or being created every year, there has been an explosion of services to allow teams to deploy their products on the internet and scale them with ease. This has created a new kind of service called cloud computing: the on-demand delivery of IT resources over the internet with pay-as-you-go model pricing.

In this book, we will deploy the backend on AWS, mostly on an **Elastic Compute Cloud** (**EC2**) instance, which is just a fancy name for a VPS. Well, actually, an AWS EC2 instance is a virtual server in Amazon's EC2 for running web applications. Let's start by creating the AWS server.

> **Important note**
>
> The following steps can work for any VPS, not just for an AWS VPS. If you can't create a VPS on AWS, you can see other solutions such as Linode, **Google Cloud Platform** (**GCP**), Azure, or IBM. They provide free credit you can use for learning about their services.

Creating an EC2 instance

Follow these steps to create an EC2 instance:

1. Make sure to be logged in to your AWS account. On the dashboard, open the EC2 console:

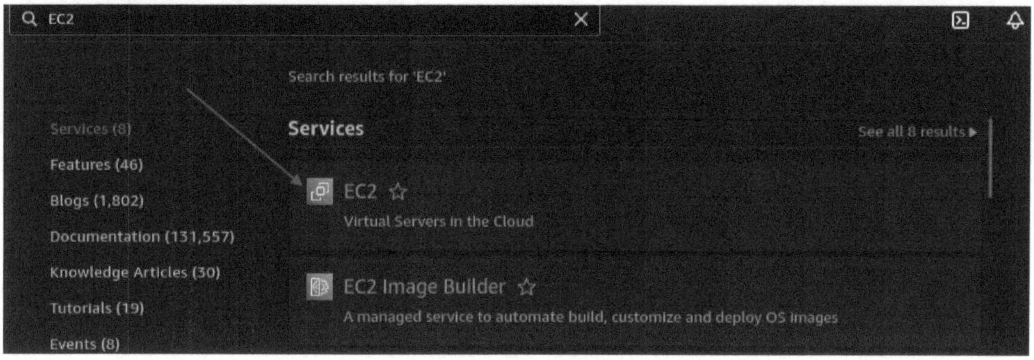

Figure 12.6 – Accessing the EC2 console

2. On the EC2 console, launch a new instance:

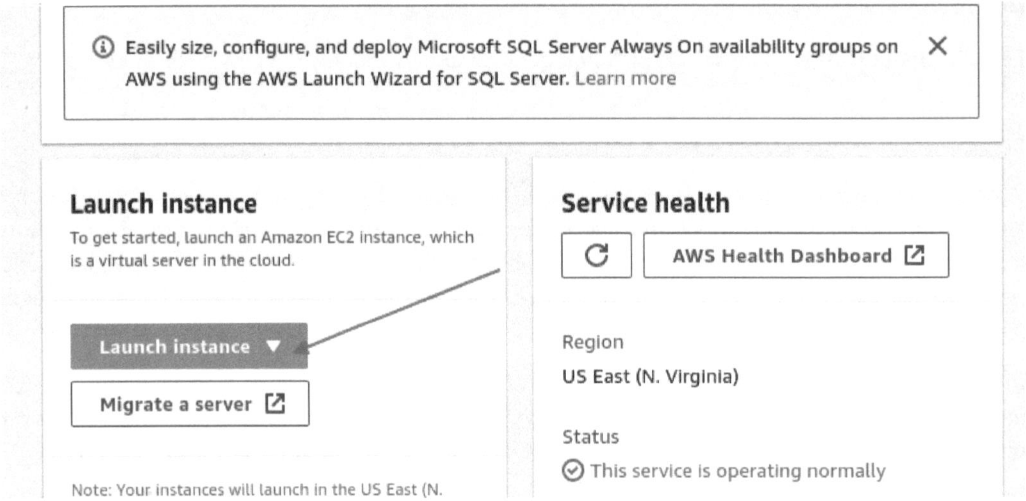

Figure 12.7 – Creating an EC2 instance

You will be shown a page where you will have to configure the instance.

3. Enter the name of the instance:

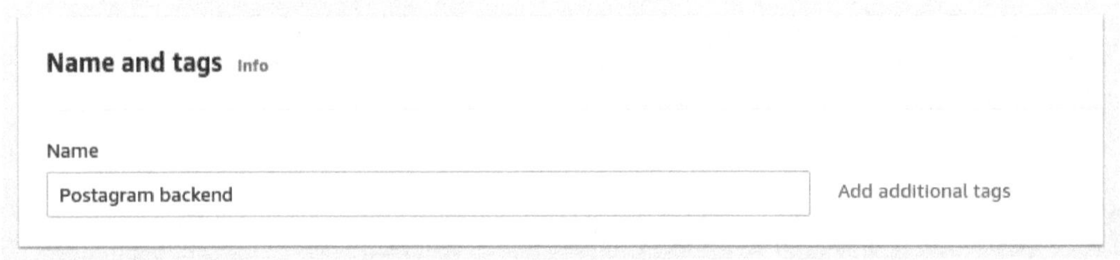

Figure 12.8 – Naming the EC2 instance

4. The next step is to choose an operating system. We will use **Ubuntu Server 22.04 LTS** for the **Amazon Machine Image** (**AMI**):

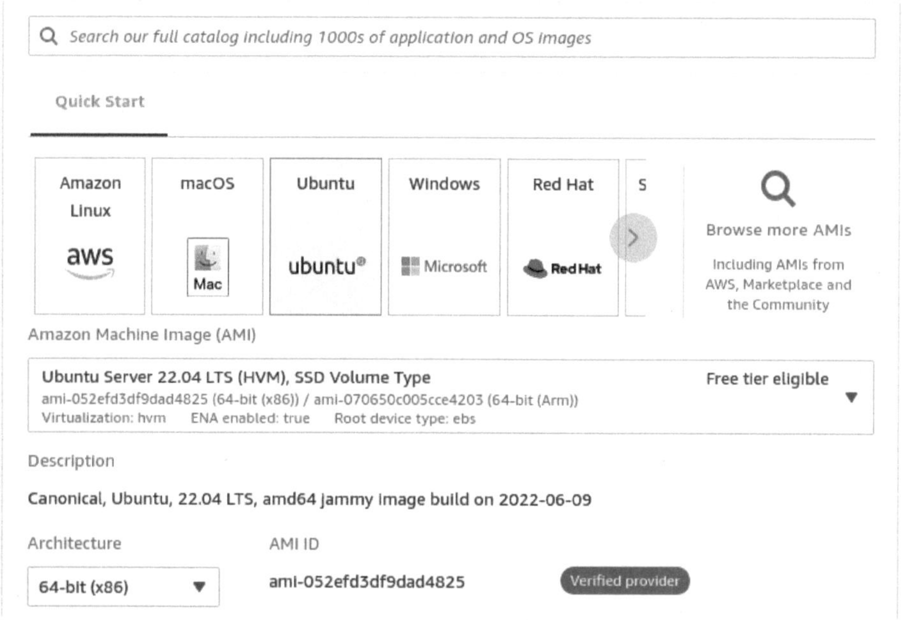

Figure 12.9 – Choosing an operating system on the EC2 instance

We are using Ubuntu here because of its security, versatility, and the policy of regular updates. However, feel free to use any other Linux distros you are familiar with.

5. And finally, you will need to set the instance type and create a pair of keys for **Secure Shell (SSH)** login. After that, you can launch the instance:

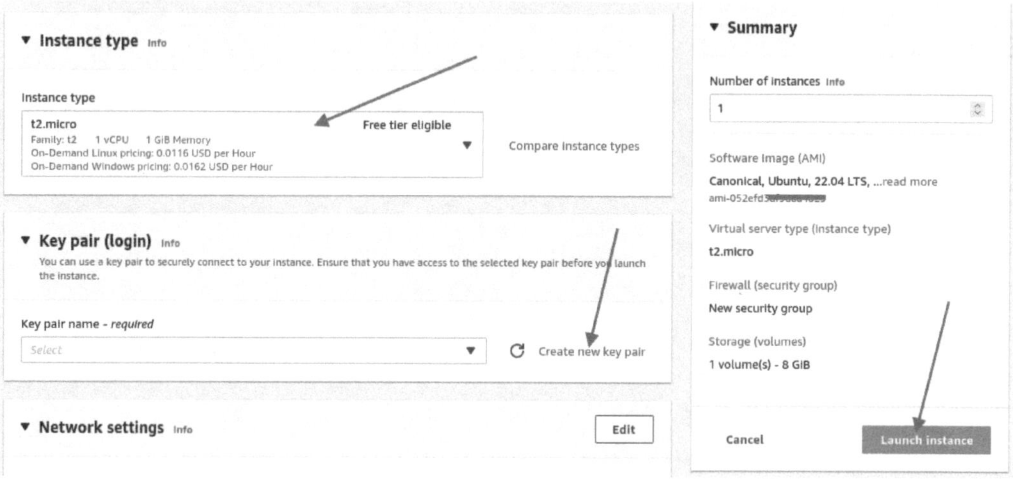

Figure 12.10 – Launching the instance

6. Wait a moment, and the instance will be created:

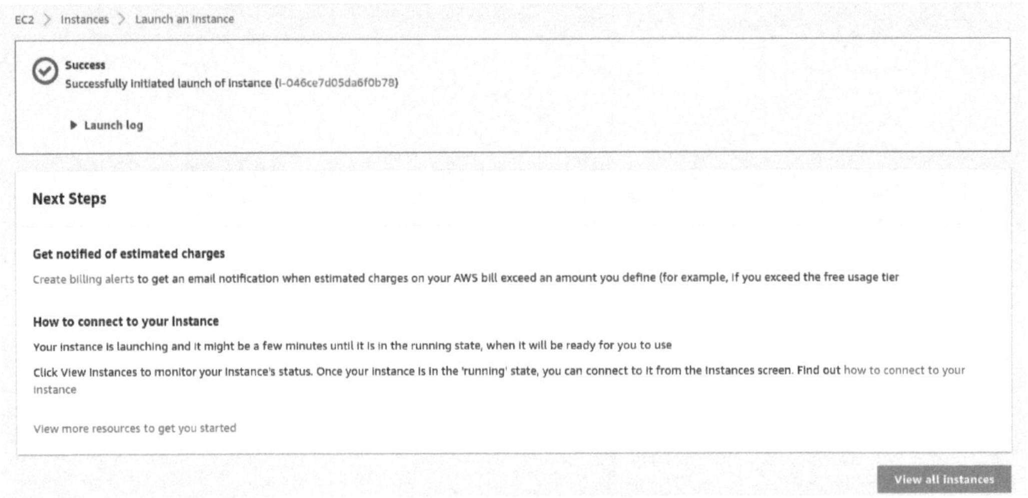

Figure 12.11 – Instance created

7. Click on the **View all instances** button, and you will see the created Postagram instance.
8. Click on the checkbox next to the name of the instance and click the **Connect** button:

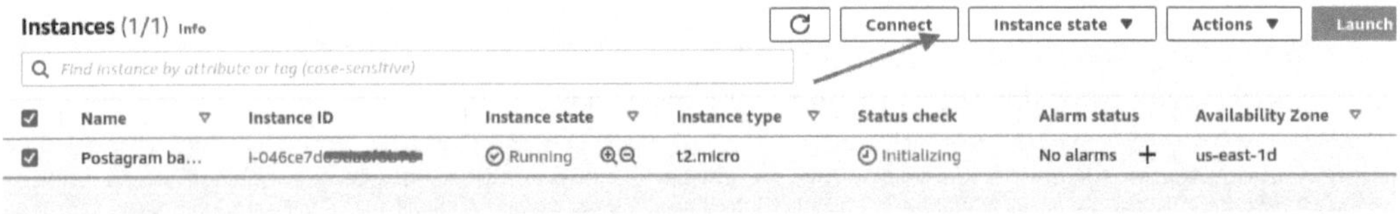

Figure 12.12 – Connecting to an EC2 instance

This will redirect you to a page with the information and steps needed to connect via SSH:

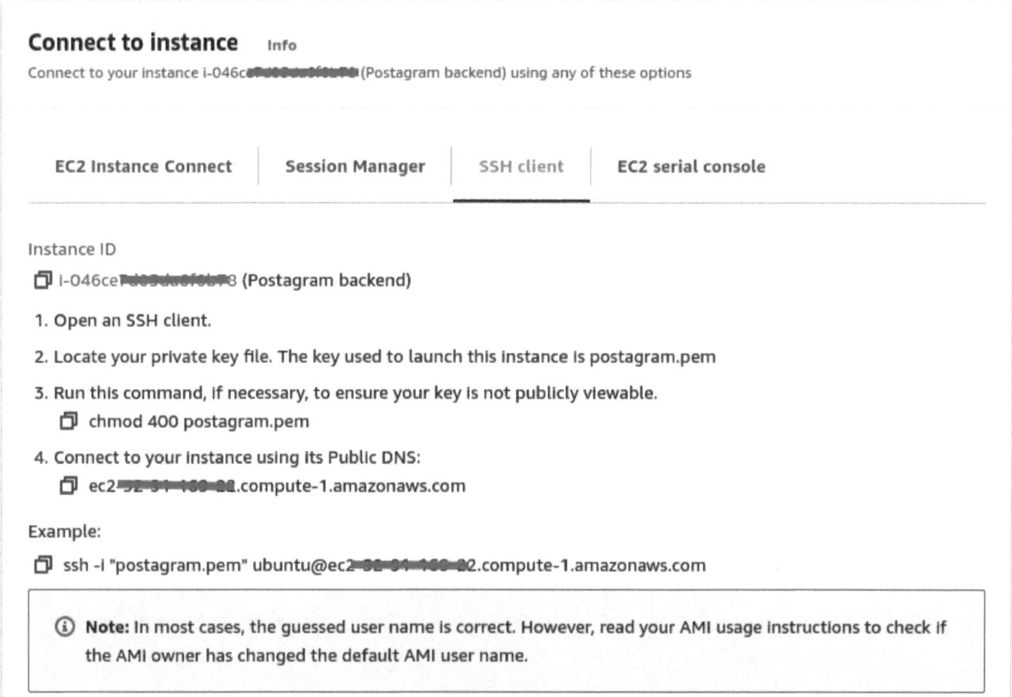

Figure 12.13 – Connecting via SSH to an EC2 instance

9. In your terminal, type the following command to connect via SSH:

```
ssh -i path/to/your_keypair.pem ec2-user@ipaddress
```

10. Once you are connected to the server, we will configure it to have a Django backend running on this machine and accessible from the internet:

```
sudo apt update
sudo apt upgrade
```

The preceding commands update the `apt` packages index of Ubuntu packages and upgrade all packages on the server.

The Django project will run on port `8000` on the machine, so we have to allow a connection to this port. By default, EC2 instances will only allow connections on ports `80` for HTTP requests, `22` for SSH connections, and—sometimes—`443` for **Secure Sockets Layer** (**SSL**) connections.

You can allow connections on port `8000` directly on the **Details** page of the created EC2 instance to access the **Security** tab on the list of tabs at the bottom of the page and click on the security setting group:

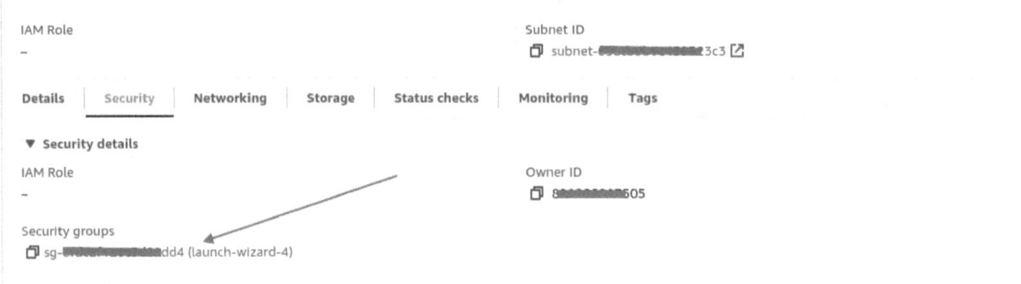

Figure 12.14 – Security tab

On the security group setting, access the **Actions** menu and click on **Edit inbound rules**. You will have access to a page where you can add a new rule, as follows:

- The type of connection is set to **Custom TCP**
- The port range is set to `8000`
- The source is set to `0.0.0.0` to indicate that all requests should be redirected to the machine on port `8000`
- And finally, add a default description to not forget why we have added this rule

Click on **Save rules** to save the changes and allow the EC2 instance to accept connections on port `8000`:

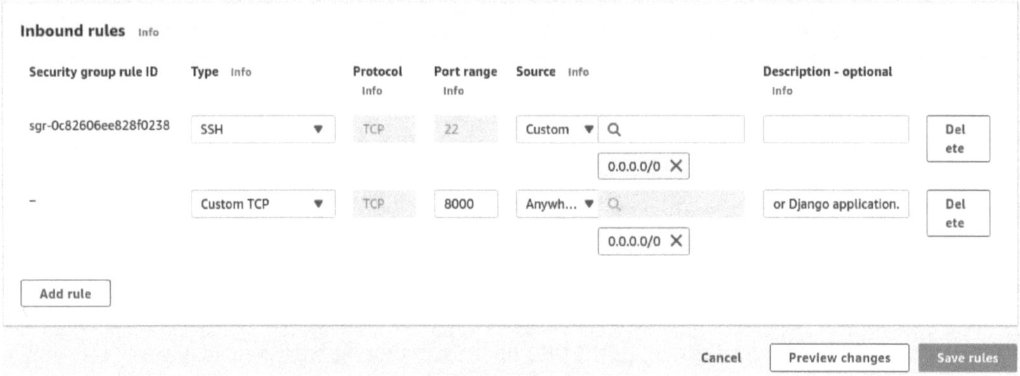

Figure 12.15 – Adding a new security rule

The server is now ready for work, and we can now run the Django backend application. Let's see the next steps in the following sections.

Configuring the server for the Django project

The source code for the Django project is hosted on GitHub. It's definitely possible to directly use `scp` to copy the code from your machine to the remote machine but let's go with Git, as it will be an important command of our workflow. On the terminal of the remote instance, enter the following command:

```
git clone your_repository_git_url
```

In my case, I am using the following repository for this project:

```
git clone https://github.com/PacktPublishing/Full-stack-Django-
and-React.git -branch chap12
```

The `git clone` command is used to get a copy of an existing repository from a remote machine on the internet or another network. The `-branch` flag is used to denote a specific branch you want to clone.

> **Important note**
>
> As I am working using the repository of the project in this book, the current code and actions done are on the `chap12` branch. In your case, if you are using your own repository, you may not have to use the `-branch` flag. Also, depending on if the GitHub repository is private or public, you will only enter your GitHub credentials if the repository is private.

The `git clone` command will clone the content of the project in a new directory. Enter the newly created directory and let's start configuring the project. We will follow most of the steps done in *Chapter 1, Creating a Django Project*, until the creation of the Django project:

1. First of all, create a virtual environment with the following command:

   ```
   python3 -m venv venv
   ```

2. And activate the virtual environment with the following command:

   ```
   source venv/bin/activate
   ```

3. Let's install the packages from the `requirements.txt` file:

   ```
   pip install -r requirements.txt
   ```

Great! The project is ready, but we need to configure a Postgres server to have the Django project running.

Postgres configuration and deployment

In *Chapter 1* of the book, *Creating a Django Project*, we configured Postgres by directly installing an executable or building the source code. On the EC2 instance, we will directly use the `apt` tool to install the Postgres server. You can follow these steps to install the Postgres server on the EC2 machine:

1. Enter the following command to install the Postgres server:

```
sudo apt install postgresql-14
```

2. Let's connect to the `psql` console and create a database:

```
sudo su postgres
psql
```

3. Great! Let's create the database with the same information on the `DATABASES` settings in the `CoreRoot/settings.py` file:

CoreRoot/settings.py

```
...
DATABASES = {
    'default': {
        'ENGINE':
          'django.db.backends.postgresql_psycopg2',
        'NAME': coredb,
        'USER': 'core',
        'PASSWORD': 'wCh29&HE&T83',
        'HOST': 'localhost',
        'PORT': '5342',
    }
}
...
```

4. Enter the following command on the `psql` console to create the `coredb` database:

```
CREATE DATABASE coredb;
```

5. To connect to the database, we need a user with a password. Execute the following command:

```
CREATE USER core WITH PASSWORD 'wCh29&HE&T83';
```

6. And the next step is to grant access to our database to the new user:

```
GRANT ALL PRIVILEGES ON DATABASE coredb TO core;
```

7. And we are nearly done. We also need to make sure this user can create a database. This will be helpful when we can run tests. To run tests, Django will configure a full environment but will also use a database:

```
GRANT CREATE PRIVILEGE TO core;
```

And we are done with the creation of the database. Next, let's connect this database to our Django project:

1. In the project directory, run the `migrate` command:

```
python manage.py migrate.
```

2. The `migrate` command should pass, and we can now start the Django server by running the following command:

```
python manage.py runserver 0.0.0.0:8000
```

3. With the Django server running, visit `http://public_ip:8000` in your web browser to access your Django project. You will have a page similar to the following figure:

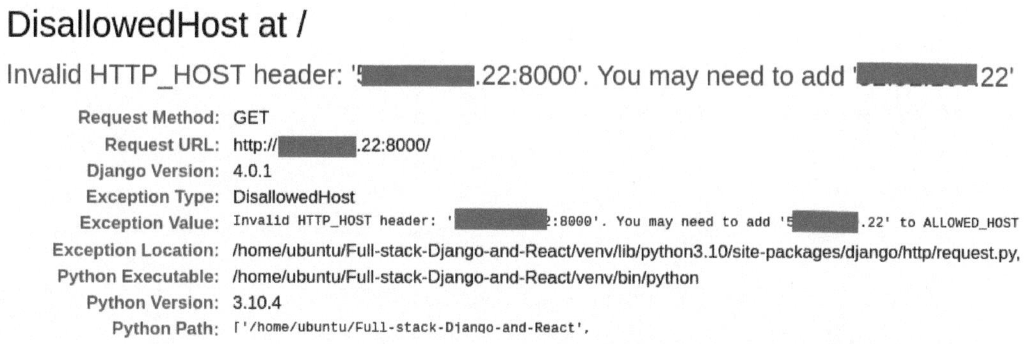

Figure 12.16 – DisallowedHost error

This is actually an error. This comes from the `ALLOWED_HOSTS` setting being empty. It is implemented by Django to prevent security vulnerabilities such as HTTP host header attacks. The `ALLOWED_HOSTS` setting contains a list of hostnames or domain names that Django can serve:

CoreRoot/settings.py

```
...
ALLOWED_HOSTS = []
...
```

4. As we are running the project from the terminal, let's modify the settings file directly on the server:

```
vim CoreRoot/settings.py
```

Or, you can use the `emacs` or `nano` command. It's up to you. The following line tells Django to accept requests from whatever is the hostname:

CoreRoot/settings.py

```
...
ALLOWED_HOSTS = ["*"]
...
```

5. Save the file and launch the server again:

```
python manage.py runserver 0.0.0.0:8000
```

6. Then, again, visit `http://public_ip:8000` in your web browser. You will see the following:

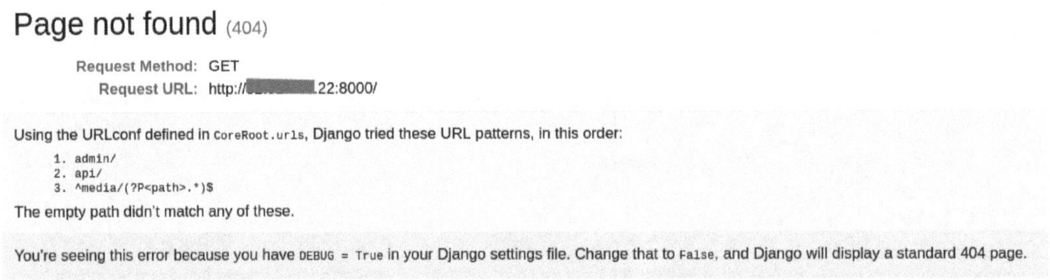

Figure 12.17 – Issues with DisallowedHost resolved

Great! The project is running fine on the internet, and you can even play with the API using an API client such as Postman or Insomnia. Congratulations! You have successfully deployed your Django application on an AWS EC2 machine.

However, we have a lot of issues (we can access debugging information directly on the internet, as in *Figure 12.17*), and we made some dangerous decisions such as not serving the API through HTTPS or not correctly setting allowed hosts throughout the deployment. Let's explore these issues in the next section.

Errors made when deploying on EC2

We have successfully deployed the Django backend on AWS. However, I decided to ignore some important and best practices for deployment so that we can have the Django server running ASAP.

Let's correct this. Let's start with the errors that Django can show us. In the terminal of the project on the remote server, run the following command:

```
python manage.py check -deploy
```

Here's the output of the preceding command:

```
System check identified some issues:
WARNINGS:
?: (security.W004) You have not set a value for the SECURE_
HSTS_SECONDS setting. If your entire site is served only over
SSL, you may want to consider setting a value and enabling HTTP
Strict Transport Security. Be sure to read the documentation
first; enabling HSTS carelessly can cause serious, irreversible
problems.
?: (security.W008) Your SECURE_SSL_REDIRECT setting is not set
to True. Unless your site should be available over both SSL and
non-SSL connections, you may want to either set this setting
True or configure a load balancer or reverse-proxy server to
redirect all connections to HTTPS.
?: (security.W009) Your SECRET_KEY has less than 50 characters,
less than 5 unique characters, or it's prefixed with 'django-
insecure-' indicating that it was generated automatically by
Django. Please generate a long and random SECRET_KEY, otherwise
many of Django's security-critical features will be vulnerable
to attack.
?: (security.W012) SESSION_COOKIE_SECURE is not set to True.
Using a secure-only session cookie makes it more difficult for
network traffic sniffers to hijack user sessions.
?: (security.W016) You have 'django.middleware.csrf.
CsrfViewMiddleware' in your MIDDLEWARE, but you have not set
CSRF_COOKIE_SECURE to True. Using a secure-only CSRF cookie
makes it more difficult for network traffic sniffers to steal
the CSRF token.
?: (security.W018) You should not have DEBUG set to True in
deployment.
System check identified 6 issues (0 silenced).
```

That's a lot of things. As we are building an API, let's focus on the security issues that concern our API:

- `SECRET_KEY`: This is an important setting in Django. It is used for all sessions, cryptographic signings, and even `PasswordReset` tokens. Having an already set value for `SECRET_KEY` can lead to dangerous security issues such as privilege escalation and remote code execution.

- DEBUG, which is set to True. That is basically why we were able to see the DisallowedHost error. Imagine an attacker going through your API, causing a 500 error, and then being able to read everything. That would be very bad.

Those are mostly the errors that Django has detected. In the last section, *Postgres configuration and deployment*, we resolved the issue of the DisallowedHost error by having Django allow whichever hostname comes in a Host header. Well, this is actually bad because it can lead to an **HTTP Host header attack**, a technique used for web cache poisoning, poisoning links in the email, and modification of sensitive operations such as password reset.

> **Important note**
> You can read more about HTTP Host header attacks at https://www.invicti.com/web-vulnerability-scanner/vulnerabilities/http-header-injection/.

There are also some issues concerning the developer experience. It's true that we have seen how to use Git and GitHub to host source code online, clone it on a remote server, and then configure it for deployment. You can repeat the same process, right? But what happens when you have to update the code for features or fixes multiple times per day? It can quickly become draining, so we need a solution for automated deployment on our EC2 server.

Also, we have Postgres and, finally, the Django project running separately. Sometimes, there might come a time when you will need to add another service to the machine. This can be done manually, but it creates an issue: the production environment starts to become different from the development environment.

It is an important habit to make sure that the development environment and the production environment are as similar as possible; this can make the reproduction of bugs easier but also the development of features predictable.

All these issues will be addressed in the next chapters. You will be introduced to environment variables, Docker, NGINX, and **continuous integration/continuous deployment** (**CI/CD**) concepts with GitHub Actions.

Summary

In this chapter, we have successfully deployed a Django application on an EC2 instance. Before deploying the Django application, we used Git to create a repository on a local machine, then created a remote repository on GitHub and pushed the changes online.

We have also learned how to configure a server for deployment manually with the installation of essential and interesting tools such as the Postgres server. We also explored the errors made when deploying the application and how we will address these errors in the following chapters.

These errors will be resolved in the next chapters, but first, we'll learn more about environment variables and Docker in the next chapter.

Questions

1. What is the usage of a Git branch?
2. What is the difference between Git and GitHub?
3. What is an HTTP Host header attack?
4. What is the use of `SECRET_KEY` in Django?

13
Dockerizing the Django Project

In the previous chapter, we learned more about software deployment, and we deployed the Django application on an AWS server. However, we came across issues such as poor preparation of the project for deployment, violation of some security issues, and deployment and development configuration.

In this chapter, we will learn how to use Docker on the Django backend and configure environment variables. We will also configure the database on a web server called **NGINX** using **Docker**. Here are the big sections of the chapter:

- What is Docker?
- Dockerizing the Django application
- Using Docker Compose for multiple containers
- Configuring environment variables in Django
- Writing NGINX configuration

Technical requirements

For this chapter, you will need to have Docker and Docker Compose installed on your machine. The Docker official documentation has a well-documented process for the installation on any OS platform. You can check it out at `https://docs.docker.com/engine/install/`.

The code written in this chapter can also be found at `https://github.com/PacktPublishing/Full-stack-Django-and-React/tree/chap13`.

What is Docker?

Before defining what **Docker** is, we must understand what a container is and its importance in today's tech ecosystem. To make it simple, a container is a standard unit of software that packages up the software and all of its required dependencies so that the software or the application can run quickly and reliably from one machine to another, whether the environment or the OS.

An interesting definition from Solomon Hykes at the 2013 *PyCon* talk is: containers are "*self-contained units of software you can deliver from a server over there to a server over there, from your laptop to EC2 to a bare-metal giant server, and it will run in the same way because it is isolated at the process level and has its own file system.*"

> **Important note**
>
> Containerization is different from virtualization. Virtualization enables teams to run multiple operating systems on the same hardware, while containerization allows teams to deploy multiple applications using the same operating system on single hardware with their own images and dependencies.

Great, right? Remember at the beginning of this book when we had to make configurations and installations depending on the OS mostly for the Python executable, the Postgres server, and different commands to create and activate a virtual environment? Using Docker, we can have a single configuration for a container, and this configuration can run the same on any machine. Docker ensures that your application can be executed in any environment. Then, we can say that Docker is a software platform for building, developing, and developing applications inside containers. It has the following advantages:

- **Minimalistic and portable**: Compared to **virtual machines** (**VMs**) that require complete copies of an OS, the application, and the dependencies, which can take a lot of space, a Docker container requires less storage because the image used comes with **megabytes** (**MB**) in size. This makes them fast to boot and easily portable even on small devices such as Raspberry Pi-embedded computers.
- **Docker containers are scalable**: Because they are lightweight, developers or DevOps can launch a lot of services based on containers and easily control the scaling using tools such as Kubernetes.
- **Docker containers are secure**: Applications inside Docker containers are running isolated from each other. Thus, a container can't check the processes running in another container.

With a better understanding of what Docker is, we can now move on to integrate Docker into the Django application.

Dockerizing the Django application

In the precedent section, we defined Docker and its advantages. In this section, we will configure Docker with the Django application. This will help you understand better how Docker works under the hood.

Adding a Docker image

A characteristic of projects that use Docker is the presence of files called **Dockerfiles** in the project. A Dockerfile is a text document that contains all the commands necessary to assemble a Docker image. A Docker image is a read-only template with instructions to create a Docker container.

Creating an image with a Dockerfile is the most popular way to go as you only need to enter the instructions you will require to set up an environment, install the package, make migrations, and a lot more. This is what makes Docker very portable. For example, in the case of our Django application, we will write the Dockerfile based on an existing image for Python `3.10` based on the popular Alpine Linux project (`https://alpinelinux.org/`). This image has been chosen because of its small size, equal to 5 MB. Inside the Dockerfile, we will also add commands to install Python and Postgres dependencies, and we will further add commands to install packages. Let's get started with the steps:

1. Start by creating a new file at the root of the Django project called `Dockerfile` and adding the first line:

Dockerfile

```
FROM python:3.10-alpine
# pull official base image
```

Most of your `Dockerfile` will start with this line. Here, we are telling Docker which image to use to build our image. The `python:3.10-alpine` image is stored in what is called a Docker registry. This is a storage and distribution system for Docker images, and you can find the most popular one online, called Docker Hub, at `https://hub.docker.com/`.

2. Next, let's set the working directory. This directory will contain the code of the running Django project:

Dockerfile

```
WORKDIR /app
```

3. As the Django application uses Postgres as a database, add the required dependencies for Postgres and Pillow to our Docker image:

Dockerfile

```
# install psycopg2 dependencies
RUN apk update \
    && apk add postgresql-dev gcc python3-dev musl-dev
    jpeg-dev zlib-dev
```

4. Then, install the Python dependencies after making a copy of the `requirements.txt` file in the `/app` working directory:

Dockerfile

```
# install python dependencies
COPY requirements.txt /app/requirements.txt
RUN pip install --upgrade pip
RUN pip install --no-cache-dir -r requirements.txt
```

5. After that, copy over the whole project itself:

Dockerfile

```
# add app
COPY . .
```

6. And finally, expose port `8000` of the container for access to the other applications or the machine, run the migrations, and start the Django server:

Dockerfile

```
EXPOSE 8000
CMD ["python", "manage.py", "migrate"]
CMD ["python", "manage.py", "runserver", "0.0.0.0:8000"]
```

The `Dockerfile` file will have the following final code:

Dockerfile

```
# pull official base image
FROM python:3.10-alpine

# set work directory
WORKDIR /app

# set environment variables
ENV PYTHONDONTWRITEBYTECODE 1
ENV PYTHONUNBUFFERED 1

# install psycopg2 dependencies
```

```
RUN apk update \
    && apk add postgresql-dev gcc python3-dev musl-dev

# install python dependencies
COPY requirements.txt /app/requirements.txt
RUN pip install --upgrade pip
RUN pip install --no-cache-dir -r requirements.txt

# copy project
COPY . .

EXPOSE 8000

CMD ["python", "manage.py", "migrate"]

CMD ["python", "manage.py", "runserver", "0.0.0.0:8000"]
```

We have just written the steps to build an image for the Django application. Let's build the image with the following command.

```
docker build -t django-postagram .
```

The preceding command uses the `Dockerfile` to build a new container image—that's why we have a dot (`.`) at the end of the command. It tells Docker to look for the `Dockerfile` in the current directory. The `-t` flag is used to tag the container image. Then, we are building an image with the `django-backend` tag using the `Dockerfile` we have written. Once the image is built, we can now run the application in the container by running the following command:

```
docker run --name django-postagram -d -p 8000:8000 django-
postagram:latest
```

Let's describe the preceding command:

- `--name` will set the name of the Docker container
- `-d` makes the image run in detached mode, meaning that it can run in the background
- `django-postagram` specifies the name of the image to use

After typing the preceding command, you can check the running container with the following command:

```
docker container ps
```

You will have a similar output:

```
(venv) koladev@koladev123xxx:~/PycharmProjects/Full-stack-Django-and-React$ docker ps
CONTAINER ID   IMAGE                     COMMAND                  CREATED         STATUS         PORTS
8c57a088fd6a   django-postagram:latest   "python manage.py ru…"   6 seconds ago   Up 6 seconds   0.0.0.0:8000->8000/tcp,
```

Figure 13.1 – Listing Docker containers on the machine

The container is created, but it looks like it's not working well. In your browser, go to `http://localhost:8000`, and you will notice that the browser returns a page with an error. Let's check the logs for the `django-postagram` container:

```
docker logs --details django-postagram
```

The command will output in the terminal what is happening inside the container. You will have a similar output to this:

```
django.db.utils.OperationalError: could not connect to server: Connection refused
        Is the server running on host "localhost" (127.0.0.1) and accepting
        TCP/IP connections on port 5432?
could not connect to server: Address not available
        Is the server running on host "localhost" (::1) and accepting
        TCP/IP connections on port 5432?
```

Figure 13.2 – Logs for the django-postagram container

Well, that's quite normal. The container is running on its own network and doesn't have direct access to the host machine network.

In the previous chapter, we added services for NGINX and Postgres and made the configurations. We need to also do the same with **Docker**; I mean, we can have two other `Dockerfiles` for NGINX and Postgres. And let's be honest: it starts to become a little bit much. Imagine adding a Flask service, a Celery service, or even another database. Depending on the *number n* of components of your system, you will need `n Dockerfiles`. This is not interesting, but thankfully, Docker provides a simple solution for that called Docker Compose. Let's explore it more.

Using Docker Compose for multiple containers

Docker Compose is a tool developed and created by the Docker team to help define configurations for multi-container applications. Using Docker Compose, we just need to create a YAML file to define the services and the command to start each service. It also supports configurations such as container name, environment setting, volume, and a lot more, and once the YAML file is written, you just need a command to build the images and spin all the services.

Let's understand the key difference between a Dockerfile and Docker Compose: a Dockerfile describes how to build the image and run the container, while Docker Compose is used to run Docker containers. At the end of the day, Docker Compose still uses Docker under the hood, and you will—most of the time—need at least a Dockerfile. Let's integrate Docker Compose into our workflow.

Writing the docker-compose.yaml file

Before writing the YAML file, we will have to make some changes to the `Dockerfile`. As we will be launching the Django server from the `docker-compose` file, we can remove the lines where we expose the port, run the migrations, and start the server. Inside the `Dockerfile`, remove the following lines of code:

Dockerfile

```
EXPOSE 8000
CMD ["python", "manage.py", "migrate"]
CMD ["python", "manage.py", "runserver", "0.0.0.0:8000"]
```

Once it's done, create a new file called `docker-compose.yaml` at the root of the project. Make sure that the `docker-compose.yaml` file and the `Dockerfile` are in the same directory. The `docker-compose.yaml` file will describe the services of the backend application. We will need to write three services:

- **NGINX**: We are using NGINX as the web server. Thankfully, there is an official image available we can use to write quick configurations.
- **Postgres**: There is also an official image available for Postgres. We will just need to add environment variables for the database user.
- `django-backend`: This is the backend application we have created. We will use the `Dockerfile` so that Docker Compose will build the image for this service.

Let's start writing the `docker-compose.yaml` file by adding the NGINX service first:

docker-compose.yaml

```
version: '3.8'

services:
 nginx:
   container_name: postagram_web
   restart: always
```

```
    image: nginx:latest
    volumes:
      - ./nginx.conf:/etc/nginx/conf.d/default.conf
      - uploads_volume:/app/uploads
    ports:
      - "80:80"
    depends_on:
      - api
```

Let's see what is going on in the preceding code because the other services will follow a similar configuration. The first line sets the file format we are using, so it is not related to Docker Compose, just to YAML.

After that, we are adding a service called `nginx`:

- `container_name` represents, well, the name of the container.
- `restart` defines the container restart policy. In this case, the container is always restarted if it fails.

  Concerning the restart policies for a container, you can also have:

  - `no`: Containers will not restart automatically
  - `on-failure[:max-retries]`: Restart the container if it exits with a nonzero exit code and provides a maximum number of attempts for the Docker daemon to restart the container
  - `unless-stopped`: Always restart the container unless it was stopped arbitrarily or by the Docker daemon
- `image`: This tells Docker Compose to use the latest NGINX image available on Docker Hub.
- `volumes` are a way of persisting data generated and used by Docker containers. If a Docker container is deleted or removed, all its content will vanish forever. This is not ideal if you have files such as logs, images, video, or anything you want to persist somewhere because every time you remove a container, this data will vanish. Here is the syntax: `/host/path:/container/path`.
- `ports`: Connection requests coming from the host port `80` are redirected to the container port `80`. Here is the syntax: `host_port:container_port`.
- `depends_on`: This tells Docker Compose to wait for some services to start before starting the service. In our case, we are waiting for the Django API to start before starting the NGINX server.

Great! Next, let's add the service configuration for the Postgres service:

docker-compose.yaml

```
db:
  container_name: postagram_db
  image: postgres:14.3-alpine
  env_file: .env
  volumes:
    - postgres_data:/var/lib/postgresql/data/
```

We have new parameters here called `env_file` which specifies the path to the environment file that will be used to create the database and the user, and set the password. Let's finally add the Django API service:

docker-compose.yaml

```
api:
  container_name: postagram_api
  build: .
  restart: always
  env_file: .env
  ports:
    - "8000:8000"
  command: >
    sh -c "python manage.py migrate --no-input && gunicorn
           CoreRoot.wsgi:application --bind 0.0.0.0:8000"
  volumes:
   - .:/app
   - uploads_volume:/app/uploads
  depends_on:
   - db
```

The build parameter in the Docker Compose file tells Docker Compose where to look for the `Dockerfile`. In our case, the `Dockerfile` is in the current directory. Docker Compose allows you to have a `command` parameter. Here, we are running migrations and starting the Django server using Gunicorn, which is new. `gunicorn` is a Python **Web Server Gateway Interface** (**WSGI**) HTTP server for Unix systems. Why use `gunicorn`? Most web applications run with an Apache server, so `gunicorn` is basically designed to run web applications built with Python.

You can install the package in your current Python environment by running the following command:

```
pip install gunicorn
```

But you will need to put the dependency in the `requirements.txt` file so that it can be preset in the Docker image:

requirements.txt

```
gunicorn==20.1.0
```

Finally, we need to declare at the end of the file the volumes used:

docker-compose.yaml

```
volumes:
  uploads_volume:
  postgres_data:
```

And we have just written a `docker-compose.yaml` file. As we are going to use environment variables in the project, let's update some variables in the `settings.py` file.

Configuring environment variables in Django

It is a bad habit to have sensitive information about your application available in the code. This is the case for the `SECRET_KEY` setting and the database settings in the `settings.py` file of the project. It is quite bad because we have pushed the code to GitHub. Let's correct this.

An environment variable is a variable whose value is set outside the running code of the program. With Python, you can read files from a `.env` file. We will use the `os` library to write the configurations. So, first, create a `.env` file at the root of the Django project and add the following content:

.env

```
SECRET_KEY=foo
DATABASE_NAME=coredb
DATABASE_USER=core
DATABASE_PASSWORD=wCh29&HE&T83
DATABASE_HOST=postagram_db
DATABASE_PORT=5432
POSTGRES_USER=core
POSTGRES_PASSWORD=wCh29&HE&T83
```

```
POSTGRES_DB=coredb
ENV=DEV
DJANGO_ALLOWED_HOSTS=127.0.0.1,localhost
```

Important note

`SECRET_KEY` is an important variable for your Django project, so you need to ensure that you have a long and complicated chain of characters as the value. You can visit `https://djecrety.ir/` to generate a new chain of characters.

The next step is to install a package to help you manage environment variables. The package is called `python-dotenv`, and it helps Python developers read environment variables from `.env` files and set them as environment variables. If you are going to run the project again on your machine, then add the package to your actual Python environment with the following command:

```
pip install python-dotenv
```

And finally, add the package to the `requirements.txt` file so that it can be installed in the Docker image. Here's a look at the `requirements.txt` file:

```
Django==4.0.1
psycopg2-binary==2.9.3
djangorestframework==3.13.1
django-filter==21.1
pillow==9.0.0
djangorestframework-simplejwt==5.0.0
drf-nested-routers==0.93.4
pytest-django==4.5.2
django-cors-headers==3.11.0
python-dotenv==0.20.0
gunicorn==20.1.0
```

Once the installation of the `python-dotenv` package is done, we need to write some code in the `CoreRoot/settings.py` file. In this file, we will import the `python-dotenv` package and modify the syntax of some settings so that it can support environment variables' reading:

CoreRoot/settings.py

```
from dotenv import load_dotenv
load_dotenv()
```

Let's rewrite the values of variables such as `SECRET_KEY`, `DEBUG`, `ALLOWED_HOSTS`, and `ENV`:

CoreRoot/settings.py

```
ENV = os.environ.get("ENV")

# SECURITY WARNING: keep the secret key used in production
secret!
SECRET_KEY = os.environ.get(
    "SECRET_KEY", default=
      "qkl+xdr8aimpf-&x(mi7)dwt^-q77aji#j*d#02-5usa32r9!y"
)

# SECURITY WARNING: don't run with debug turned on in
production!
DEBUG = False if ENV == "PROD" else True

ALLOWED_HOSTS = os.environ.get("DJANGO_ALLOWED_HOSTS",
default="*").split(",")
```

The `os` package provides an object to retrieve environment variables from the user machine. After `python-dotenv` has forced the loading of the environment variables, we use `os.environ` to read the values from the `.env` file. Let's finally add the configuration for the `DATABASES` setting:

CoreRoot/settings.py

```
DATABASES = {
    "default": {
        "ENGINE": "django.db.backends.postgresql_psycopg2",
        "NAME": os.getenv("DATABASE_NAME", "coredb"),
        "USER": os.getenv("DATABASE_USER", "core"),
        "PASSWORD": os.getenv("DATABASE_PASSWORD",
                                      "wCh29&HE&T83"),
        "HOST": os.environ.get("DATABASE_HOST",
                                      "localhost"),
        "PORT": os.getenv("DATABASE_PORT", "5432"),
    }
}
```

Great! We are done configuring the environment variables in the `settings.py` file. We can now move on to write the configurations for NGINX.

Writing NGINX configuration

NGINX requires some configuration from our side. If there is a request on the HTTP port of the machine (by default `80`), it should redirect the requests to port `8000` of the running Django application. Put simply, we will write a reverse proxy. A proxy is an intermediary process that takes an HTTP request from a client, passes the request to one or many other servers, waits for a response from those servers, and sends back a response to the client.

By using this process, we can forward a request on the HTTP port `80` to port `8000` of the Django server.

At the root of the project, create a new file called `nginx.conf`. Then, let's define the upstream server where HTTP requests will be redirected to:

nginx.conf

```
upstream webapp {
   server postagram_api:8000;
}
```

The preceding code follows the simple syntax shown next:

```
upstream upstream_name {
   server host:PORT;
}
```

> **Important note**
>
> Docker allows you to refer to the container's host with the `defined` container name. In the NGINX file, we are using `postagram_api` instead of the IP address of the container, which can change, and for the database, we are using `postagram_db`.

The next step is to declare the configuration for the HTTP server:

nginx.conf

```
server {
   listen 80;
   server_name localhost;
```

```
    location / {
        proxy_pass http://webapp;
        proxy_set_header X-Forwarded-For
          $proxy_add_x_forwarded_for;
        proxy_set_header Host $host;
        proxy_redirect off;
    }

    location /media/ {
     alias /app/uploads/;
    }
}
```

In the server configuration, we first set the port of the server. In the preceding code, we are using port 80. Next, we are defining locations. A location in NGINX is a block that tells NGINX how to process the request from a certain URL:

- A request on the / URL is redirected to the web app upstream
- A request on the /media/ URL is redirected to the uploads folder to serve files

With the NGINX configuration ready, we can now launch the containers.

Launching the Docker containers

Let's launch the Docker containers. As we are now using Docker Compose to orchestrate containers, let's use the following command to build and start the containers:

```
docker compose up -d -build
```

This command will spin up all the containers defined in the docker-compose.yaml file. Let's describe the command options:

- up: This builds, recreates, and starts containers
- -d: This is used to detach, meaning that we are running the containers in the background
- –build: This flag tells Docker Compose to build the images before starting the containers

After the build is done, open your browser at `http://localhost`, and you should see the following:

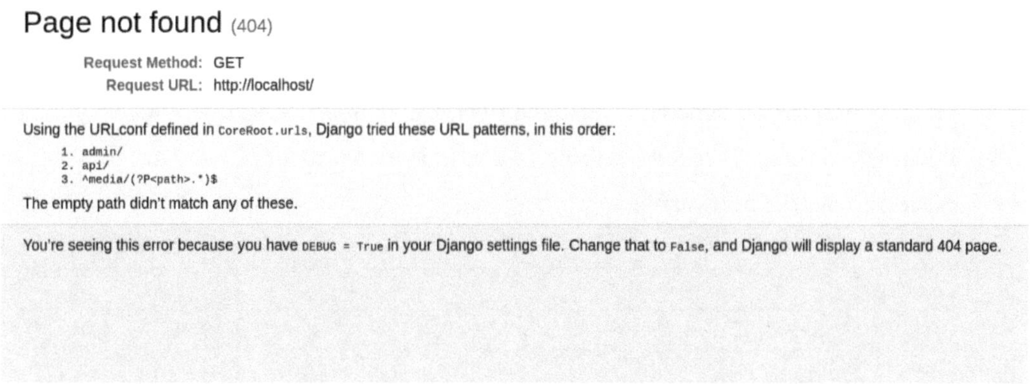

Figure 13.3 – Dockerized Django application

We have successfully containerized the Django application using Docker. It is also possible to execute commands inside containers, and right now, we can start by running a test suite inside the `postagram_api` container:

```
docker compose exec -T api pytest
```

The syntax to execute a command in a Docker container is to first call the `exec` command followed by the `-T` parameter to disable `pseudo-tty` allocation. This means that the command being run inside the container will not be attached to a terminal. Finally, you can add the container service name, followed by the command you want to execute in the container.

We are one step closer to the deployment of AWS using **Docker**, but we need to automate it. In the next chapter, we will configure the project with GitHub Actions to automate deployment on the AWS server.

Summary

In this chapter, we have learned how to dockerize a Django application. We started by looking into Docker and its use in the development of modern applications. We also learned how to build a Docker image and run a container using this image—this introduced us to some limitations of Dockerization using Dockerfiles. This led us to learn more about Docker Compose and how it can help us manage multiple containers with just one configuration file. This in turn directed us to configure a database and an NGINX web server with Docker to launch the Postagram API.

In the next chapter, we will configure the project for automatic deployment on AWS but also carry out regression checks using the tests we have written.

Questions

1. What is Docker?
2. What is Docker Compose?
3. What is the difference between Docker and Docker Compose?
4. What is the difference between containerization and virtualization?
5. What is an environment variable?

14
Automating Deployment on AWS

In the previous chapter, we successfully deployed the Django application on an EC2 instance. However, most of the deployment is done manually, and we don't check for regression when pushing a new version of the application. Interestingly, all the deploying can be automated using GitHub Actions.

In this chapter, we will use GitHub Actions to automatically deploy on an AWS EC2 instance so that you don't have to do it manually. We will explore how to write a configuration file that will run tests on the code to avoid regressions, and finally connect via **Secure Socket Shell** (**SSH**) to a server and execute the script to pull and build the recent version of the code and up the container. To recapitulate, we will cover the following topics:

- Explaining **continuous integration and continuous deployment** (**CI/CD**)
- Defining the CI/CD workflow
- What is GitHub Actions?
- Configuring the backend for automated deployment

Technical requirements

The code for this chapter can be found at `https://github.com/PacktPublishing/Full-stack-Django-and-React/tree/chap14`. If you are using a Windows machine, ensure that you have the OpenSSH client installed on your machine as we will generate SSH key pairs.

Explaining CI/CD

Before going deeper into GitHub Actions, we must understand the terms *CI* and *CD*. In this section, we will understand each term and explain the differences.

CI

CI is a practice of automating the integration of code changes from multiple collaborators into a single project. It also concerns the ability to reliably release changes made to an application at any time. Without CI, we should have to manually coordinate the deployment, the integration of changes into an application, and security and regression checks.

Here's a typical CI workflow:

1. A developer creates a new branch from the main branch, makes changes, commits, and then pushes it to the branch.
2. When the push is done, the code is built, and then automated tests are run.
3. If the automated tests fail, the developer team is notified, and the next steps (usually deployment) are canceled. If the tests succeed, then the code is ready to be deployed in a staging or production environment.

You can find many tools for CI pipeline configurations. You have tools such as GitHub Actions, Semaphore, Travis CI, and a lot more. In this book, we will use GitHub Actions to build the CI pipeline, and if the CI pipeline passes, we can deploy it on AWS. Let's now learn more about CD.

CD

CD is related to CI but most of the time represents the next step after a successful CI pipeline passes. The quality of the CI pipeline (builds and tests) will determine the quality of the releases made. With CD, the software is automatically deployed to a staging or production environment once it passes the CI step.

An example of a CD pipeline could look like this:

1. A developer writes a branch, makes changes and pushes the changes, and then creates a merge request.
2. Tests and builds are done to make sure there is no regression.
3. The code is reviewed by another developer, and if the review is done, the merge request is validated and then another suite of tests and builds are done.
4. After that, the changes are deployed to a staging or production environment.

GitHub Actions and the other tools mentioned for CI also support CD. With a better understanding of CI and CD, let's define the workflow that we will configure for the backend.

> **Important note**
>
> You will also hear about *continuous delivery* if you are diving deeper into CI/CD; it is a further extension of *continuous deployment*. Continuous deployment focuses on the deployment of the servers while continuous delivery focuses on the release and release strategy.

Defining the CI/CD workflow

Before deploying an application as we did in the previous chapter, we need to write off the steps we will follow, along with the tools needed for the deployment. In this chapter, we will automate the deployment of the backend on AWS. Basically, each time we have a push made on the main branch of the repository, the code should be updated on the server and the containers should be updated and restarted.

Again, let's define the flow, as follows:

1. A push is made on the principal branch of the server.
2. Docker containers are built and started to run tests. If the tests fail, the following steps are ignored.
3. We connect via SSH to the server and run a script to pull the new changes from the remote repository, build the containers, and restart the services using `docker-compose`.

The following diagram illustrates a typical CI/CD workflow:

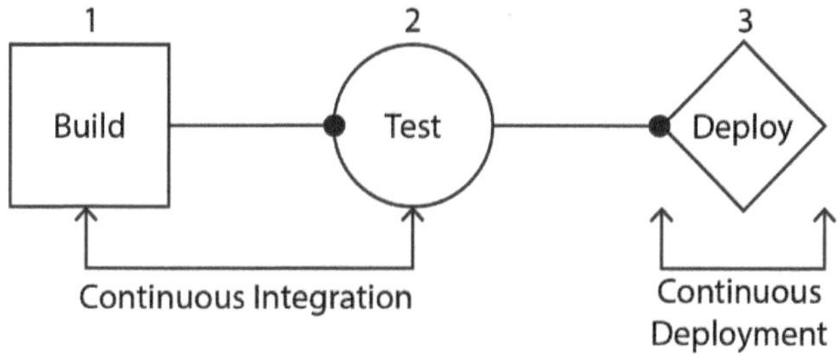

Figure 14.1 – CI/CD workflow

That is a lot of things to do manually, and thankfully, GitHub provides an interesting feature called GitHub Actions. Now that we have a better idea about the deployment strategy, let's explore this feature more.

What is GitHub Actions?

GitHub Actions is a service built and developed by GitHub for automating builds, testing, and deployment pipelines. Using GitHub Actions, we can easily implement the CI/CD workflow shown in *Figure 14.1*. Before continuing, make sure that your project is hosted on GitHub.

GitHub Actions configurations are made in a file that must be stored in a dedicated directory in the repository called `.github/workflows`. For a better workflow, we will also use GitHub secrets to store deployment information such as the IP address of the server, the SSH passphrase, and the server username. Let's start by understanding how to write a GitHub Actions workflow file.

How to write a GitHub Actions workflow file

Workflow files are stored in a dedicated directory called `.github/workflows`. The syntax used for these files is YAML syntax, hence workflow files have the `.yml` extension.

Let's dive deeper into the syntax of a workflow file:

- `name`: This represents the name of the workflow. This name is set by placing the following line at the beginning of the file:

  ```
  name: Name of the Workflow
  ```

- `on`: This specifies the events that will trigger the workflow automatically. An example of an event is a push, a pull request, or a fork:

  ```
  on: push
  ```

- `jobs`: This specifies the actions that the workflow will perform. You can have multiple jobs and even have some jobs depending on each other:

  ```
  jobs:
   build-test:
     runs-on: ubuntu-latest
     steps:
     - uses: actions/checkout@v2
     - name: Listing files in a directory
       run: ls -a
  ```

  In our GitHub Actions workflow, we will have two jobs:

  - A job named `build-test` to build the Docker containers and run the tests inside those containers
  - A job named `deploy` to deploy the application to the AWS server

The deployment of the application will depend on the failure or success of the `build-test` job. It's a good way to prevent code from failing and crashing in the production environment. Now that we understand the GitHub Actions workflow, YAML syntax, and the jobs we want to write for our workflow, let's write the GitHub Actions file and configure the server for automatic deployment.

Configuring the backend for automated deployment

In the previous sections, we discussed more about the syntax of a GitHub Actions file and the jobs we must write to add CI and CD to the Django application. Let's write the GitHub Action file and configure the backend for automatic deployment.

Adding the GitHub Actions file

At the root of the project, create a directory called `.github`, and inside this directory create another directory called `workflows`. Inside the `workflows` directory, create a file called `ci-cd.yml`. This file will contain the YAML configuration for the GitHub action. Let's start by defining the name and the events that will trigger the running of the workflow:

.github/workflows/ci-cd.yml

```
name: Build, Test and Deploy Postagram
on:
 push:
   branches: [ main ]
```

The workflow will run every time there is a push on the main branch. Let's go on to write a `build-test` job. For this job, we will follow three steps:

1. Injecting environment variables into a file. Docker will need a `.env` file to build the images and start the containers. We'll inject dummy environment variables into the Ubuntu environment.
2. After that, we will build the containers.
3. And finally, we run the tests on the `api` container.

Let's get started with the steps:

1. Let's start by writing the job and injecting the environment variables:

.github/workflows/ci-cd.yml

```
build-test:
 runs-on: ubuntu-latest
 steps:
 - uses: actions/checkout@v2
 - name: Injecting env vars
   run: |
     echo "SECRET_KEY=test_foo
```

```
        DATABASE_NAME=test_coredb
        DATABASE_USER=test_core
        DATABASE_PASSWORD=12345678
        DATABASE_HOST=test_postagram_db
        DATABASE_PORT=5432
        POSTGRES_USER=test_core
        POSTGRES_PASSWORD=12345678
        POSTGRES_DB=test_coredb
        ENV=TESTING
        DJANGO_ALLOWED_HOSTS=127.0.0.1,localhost
         " >> .env
```

The tests will probably fail because we haven't defined the Github Secret called `TEST_SECRETS`.

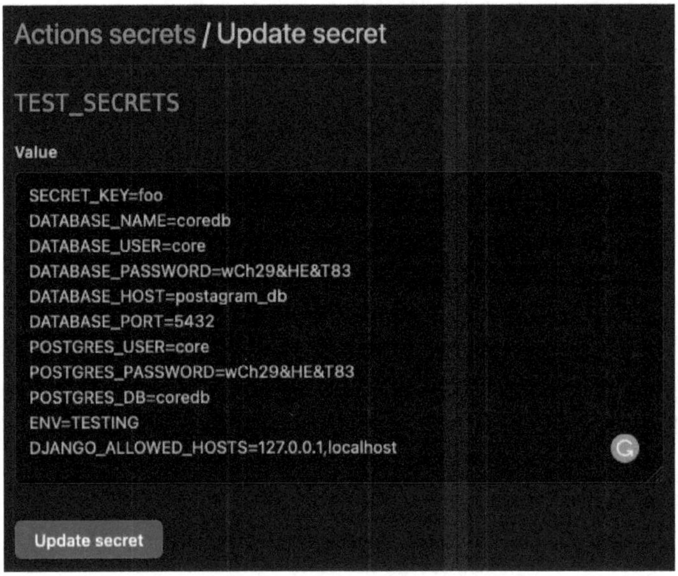

Figure 14.2 – Testing Github secrets

2. Next, let's add the command to build the containers:

.github/workflows/ci-cd.yml

```
  - name: Building containers
    run: |
      docker-compose up -d --build
```

3. And finally, let's run the `pytest` command in the `api` container:

.github/workflows/ci-cd.yml

```
- name: Running Tests
  run: |
    docker-compose exec -T api pytest
```

Great! We have the first job of the workflow fully written.

4. Let's push the code by running the following command and see how it runs on the GitHub side:

```
git push
```

5. Go to GitHub to check your repository. You will see an orange badge on the details of the repository, meaning that the workflow is running:

Figure 14.3 – Running GitHub Actions

6. Click on the orange badge to have more details about the running workflows. The workflow should pass, and you will have a green status:

Figure 14.4 – Successful GitHub Action job

Great! We have the `build-test` job running successfully, which means that our code can be deployed in a production environment. Before writing the `deploy` job, let's configure the server first for automatic deployment.

Configuring the EC2 instance

It's time to go back to the EC2 instance and make some configurations to ease the automatic deployment. Here's the list of tasks to do so that GitHub Actions can automatically handle the deployment for us:

- Generate a pair of SSH keys (private and public keys) with a passphrase.
- Add the public key to `authorized_keys` on the server.

- Add the private key to GitHub Secrets to reuse it for the SSH connection.
- Register the username used on the OS of the EC2 instance, the IP address, and the SSH passphrase to GitHub Secrets.
- Add a deploying script on the server. Basically, the script will pull code from GitHub, check for changes, and eventually build and rerun the containers.
- Wrap everything and add the `deploy` job.

This looks like a lot of steps, but here's the good thing: you just need to do that once. Let's start by generating SSH credentials.

Generating SSH credentials

The best practice for generating SSH keys is to generate the keys on the local machine and not the remote machine. In the next lines, we will use terminal commands. If you are working on a Windows machine, make sure you have the OpenSSH client installed. The following commands are executed on a Linux machine. Let's get started with the steps:

1. Open the terminal and enter the following command to generate an RSA key pair:

```
ssh-keygen -t rsa -b 4096 -C "your_email@example.com"
```

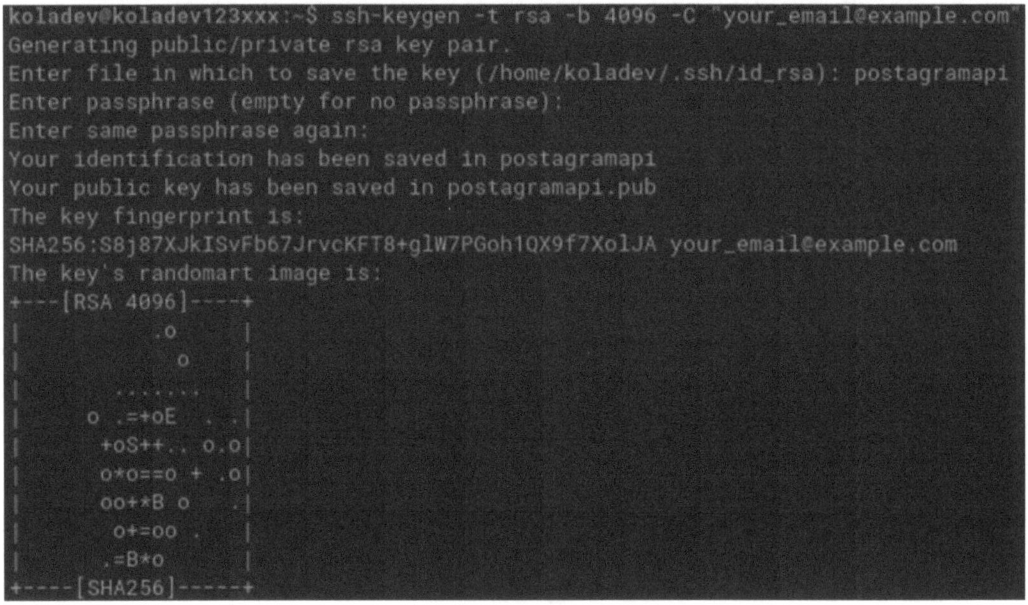

Figure 14.5 – Generating SSH keys

2. Next, copy the content of the public key and add it to the `.ssh/authorized_keys` file of the remote EC2 instance. You can just do a copy and paste using the mouse, or you can type the following command:

```
cat .ssh/postagramapi.pub | ssh username@hostname_or_
ipaddress 'cat >> .ssh/authorized_keys'
```

3. Then, copy the content of the private key and add it to GitHub Secrets:

Figure 14.6 – Registering the private key into GitHub Secrets

You also need to do the same for the passphrase, EC2 server IP address, and username for the OS of the EC2 machine:

Figure 14.7 – Repository secrets

Great! We have the secrets configured on the repository; we can now write the `deploy` job on the GitHub action.

Adding a deploying script

The benefit of using GitHub Actions is that you can already find preconfigured GitHub Actions on GitHub Marketplace and just use them instead of reinventing the wheel. For the deployment, we will use the `ssh-action` GitHub action, which is developed to allow developers to execute remote commands via SSH. This perfectly fits our needs.

Let's write the `deploy` job inside our GitHub action workflow and write a deployment script on the EC2 instance:

1. Inside the `.github/workflows/ci-cd.yml` file, add the following code at the end of the file:

.github/workflows/ci-cd.yml

```
deploy:
  name: Deploying on EC2 via SSH
  if: ${{ github.event_name == 'push' }}
  needs: [build-test]
  runs-on: ubuntu-latest
  steps:
  - name: Deploying Application on EC2
    uses: appleboy/ssh-action@master
    with:
      host: ${{ secrets.SSH_EC2_IP }}
      username: ${{ secrets.SSH_EC2_USER }}
      key: ${{ secrets.SSH_PRIVATE_KEY }}
      passphrase: ${{ secrets.SSH_PASSPHRASE }}
      script: |
        cd ~/.scripts
        ./docker-ec2-deploy.sh
```

The script run on the EC2 instance is the execution of a file called `docker-ec2-deploy.sh`. This file will contain Bash code for pulling code from the GitHub repository and building the containers.

Let's connect to the EC2 instance and add the `docker-ec2-deploy.sh` code.

2. In the home directory, create a file called `docker-ec2-deploy.sh`. The process for deployment using Git and Docker will follow these steps:

   I. We must ensure that there are effective changes in the GitHub repository to continue with building and running the containers. It will be a waste of resources and memory to rebuild the containers if the Git pull hasn't brought new changes. Here's how we can check this:

```
#!/usr/bin/env bash

TARGET='main'

cd ~/api || exit

ACTION_COLOR='\033[1;90m'
NO_COLOR='\033[0m'

echo -e ${ACTION_COLOR} Checking if we are on the target
branch
BRANCH=$(git rev-parse --abbrev-ref HEAD)
if [ "$BRANCH" != ${TARGET} ]
then
   exit 0
fi
```

   II. Next step, we will do a `git fetch` command to download content from the GitHub repository:

```
# Checking if the repository is up to date.

git fetch
HEAD_HASH=$(git rev-parse HEAD)
UPSTREAM_HASH=$(git rev-parse ${TARGET}@{upstream})

if [ "$HEAD_HASH" == "$UPSTREAM_HASH" ]
then
   echo -e "${FINISHED}"The current branch is up to date
with origin/${TARGET}."${NO_COLOR}"
     exit 0
fi
```

Once this is done, we will then check the repository is up to date by comparing the `HEAD` hash and the `UPSTREAM` hash. If they are the same, then the repository is up to date.

III. If the `HEAD` and the `UPSTREAM` hashes are not the same, we pull the latest changes, build the containers, and run the containers:

```
# If there are new changes, we pull these changes.

git pull origin main;

# We can now build and start the containers

docker compose up -d --build

exit 0;
```

Great! We can now give execution permission to the script:

```
chmod +x docker-ec2-deploy.sh
```

And we are done. You can push the changes made on the GitHub workflow and the automatic deployment job will start.

> **Important note**
>
> Depending on the type of repository (private or public), you might need to enter your GitHub credentials on every remote git command executed such as `git push` or `git pull` for example. Ensure you have your credentials configured using SSH or HTTPS. You can check how to do it `https://docs.github.com/en/authentication/keeping-your-account-and-data-secure/creating-a-personal-access-token`

Ensure to have a `.env` file at the root of the project in the AWS server. Here is an example of a `.env` file you can use for deployment. Don't forget to change the values of database credentials or secret keys:

```
SECRET_KEY=foo
DATABASE_NAME=coredb
DATABASE_USER=core
DATABASE_PASSWORD=wCh29&HE&T83
DATABASE_HOST=localhost
DATABASE_PORT=5432
POSTGRES_USER=core
```

```
POSTGRES_PASSWORD=wCh29&HE&T83
POSTGRES_DB=coredb
ENV=PROD
DJANGO_ALLOWED_HOSTS=EC2_IP_ADDRESS,EC2_INSTANCE_URL
```

Ensure to replace the `EC2_IP_ADDRESS` and the `EC2_INSTANCE_URL` with the values of your EC2 instance. You will also need to allow TCP connections on port **80** to allow HTTP requests on the EC2 instances for the whole configuration to work.

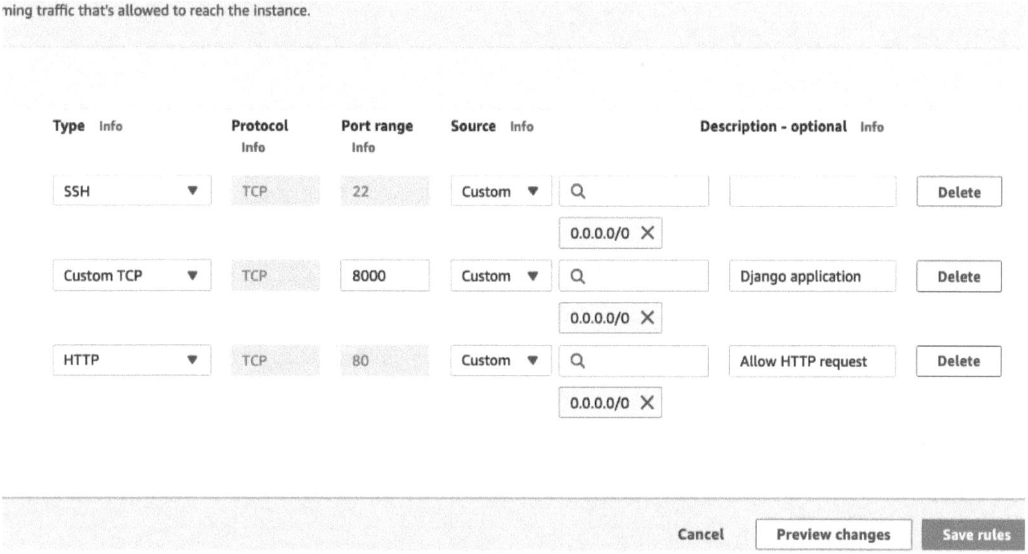

Figure 14.8 – Allowing HTTP requests

You can also remove the **8000** configurations as NGINX handles the redirection of HTTP requests to **0.0.0.0:8000** automatically.

With the concept of CI/CD understood and GitHub Actions explained and written, you have all the tools you need now to automate deployment on EC2 instances and any server. Now that the backend is deployed, we can move on to deploying the React frontend, not on an EC2 instance but on AWS **Simple Storage Service (S3)**.

Summary

In this chapter, we have finally automated the deployment of the Django application on AWS using GitHub Actions. We have explored the concepts of CI and CD and how GitHub Actions allow the configuration of such concepts.

We have written a GitHub action file with jobs to build and run the test suites, and if these steps are successful, we run the `deploy` job, which is just connecting to the EC2 instance, and run a script to pull changes, build new images, and run the containers.

In the next chapter, we will learn how to deploy the React application using a service such as AWS S3.

Questions

1. What is the difference between CI and CD?
2. What are GitHub Actions?
3. What is continuous delivery?

15
Deploying Our React App on AWS

In the previous chapter, we automated the deployment of the Django application using GitHub Actions and by making some configurations on the AWS EC2 instance. The Postagram API is live and now we must deploy the React application to have the full Postagram application available on the internet.

In this chapter, we will deploy the React application using AWS **Simple Storage Service** (**S3**) and automate the deployment using GitHub Actions. We will cover the following topics:

- Deployment of React applications
- Deploying on AWS S3
- Automating deployment with GitHub Actions

Technical requirements

For this chapter, you will need to have an account on AWS. You will also need to create an **Identity and Access Management** (**IAM**) user and save the credentials. You can do this by following the official documentation at `https://docs.aws.amazon.com/IAM/latest/UserGuide/id_users_create.html#id_users_create_cliwpsapi`. You can find the code for this chapter at `https://github.com/PacktPublishing/Full-stack-Django-and-React/tree/chap15`.

Deployment of React applications

A React application is built using JavaScript and JSX. However, to make the application accessible on the internet, we need a version of the application that a browser can interpret and understand, basically having an application with HTML, CSS, and JavaScript.

In development mode, React provides an environment for detecting warnings and tools to detect and fix problems in the application and eliminate potential issues. This adds extra code to the project, increasing the bundle size and resulting in a bigger and slower application.

It is crucial to only deploy production-built applications on the internet because of the **user experience** (**UX**). According to Google studies, *53% of users leave a website if it takes more than 3 seconds to load.* Thus, we must build the React application we created and deploy the production version.

What is a production build?

In development, the React application runs in development mode or local mode. This is where you can see all the warnings and the traceback in case your code crashes. The production mode requires the developers to build the application. This build minifies the code, optimizes the assets (image, CSS files, and so on), produces lighter source maps, and suppresses the warning messages displayed in development mode.

Therefore, the bundle size of the application is drastically reduced, and this improves page load speed. In this chapter, we will build a production-ready application and deploy it on AWS S3 as a static website.

Deploying on AWS S3

AWS S3 is one of the most popular services of AWS. It is a cloud-based storage service providing high performance, availability, reliability, security, and a ridiculous potential for scaling. AWS S3 is mostly used to store static assets so that they are effectively distributed to the internet, and because of the distribution characteristic, AWS S3 is suitable for hosting static websites.

In this chapter, we will create an S3 bucket, upload the content of the built React application, and allow public access from the internet. An **S3 bucket** is just a public storage resource available in AWS that is like an online folder where you can store objects (like a folder on your Google Drive). In the next section, we will create a production-ready version of the React application.

Creating a build of Postagram

We can create a build of the React application with just one command:

```
Yarn build
```

The `yarn build` command creates a bundle of static files of a React application. This bundle is optimized enough to go into production. The production Postagram application will use the online version of the API. This means we need to make some readjustments in the React code, mainly concerning the API URLs used in the code.

In *Part 2* of this book, *Build Reactive UI with React*, we built the React application using data from the localhost server at port `8000`. In this chapter, it won't be the case, and we will take the occasion to add environment variables to the React application. Integrating environment variables into a React application is straightforward. Let's configure the environment variables in the Postagram React application.

Adding environment variables and building the application

According to the documentation of *Create React App* regarding environment variables (`https://create-react-app.dev/docs/adding-custom-environment-variables/`),

> *"Your project can consume variables declared in your environment as if they were declared locally in your JS files. By default, you will have NODE_ENV defined for you, and any other environment variables starting with REACT_APP_".*

To access the values of the environment variables, we will use `process.env.REACT_APP_VALUE` syntax because these environment variables are defined on `process.env`.

At the root of the React project, create a file called `.env`. Inside this file, add the following content and the name of the API URL you have deployed on the EC2 AWS server:

```
REACT_APP_API_URL=https://name_of_EC2_instance.compute-1.
amazonaws.com/api
```

You then need to modify some pieces of code at `src/helpers/axios.js` and `src/hooks/user.actions.js`. We must update the `baseURL` variable to read the values from the `.env` file:

src/hooks/user.actions.js

```
function useUserActions() {
  const navigate = useNavigate();
  const baseURL = process.env.REACT_APP_API_URL;

  return {
    login,
    register,
    logout,
    edit,
  };
```

And we do the same on the axios.js file:

src/helpers/axios.js

```
const axiosService = axios.create({
 baseURL: process.env.REACT_APP_API_URL,
 headers: {
   "Content-Type": "application/json",
 },
});
...
const refreshAuthLogic = async (failedRequest) => {
 return axios
   .post(
     "/auth/refresh/",
     {
       refresh: getRefreshToken(),
     },
     {
       baseURL: process.env.REACT_APP_API_URL,
...
```

Great! The application can be built now. Run the following command:

```
yarn build
```

You will have a similar result to this:

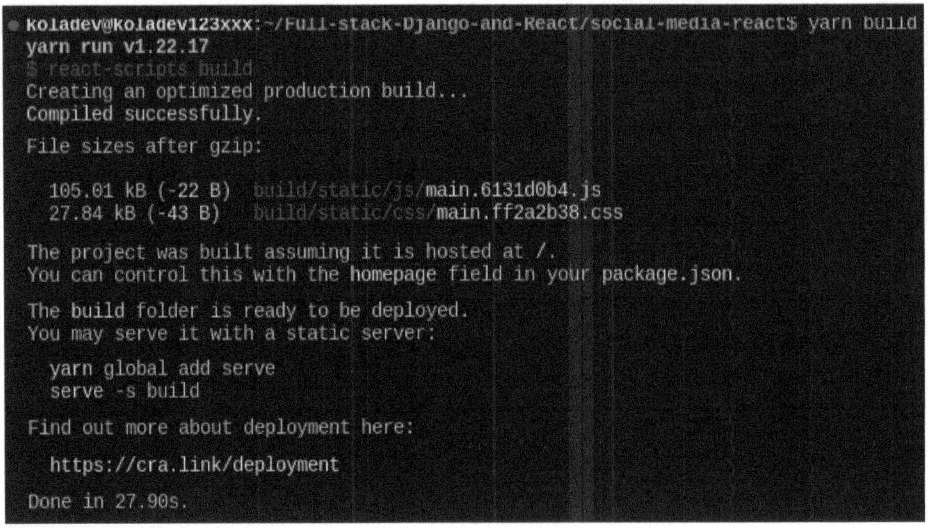

Figure 15.1 – Output of yarn build command

The build is available in the newly created `build` directory, where you will find the following content:

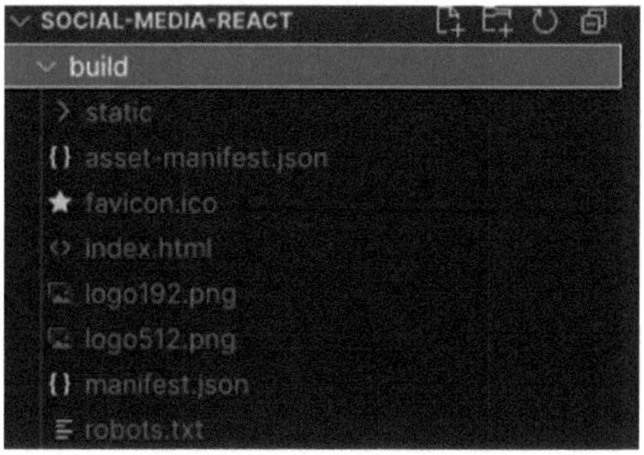

Figure 15.2 – build directory

With a production-ready React application, we can then deploy the application on S3. Next, let's create an S3 bucket and upload the files and folders.

Deploying the React application on S3

We have a build-ready version of the application and an optimized version for production. Before deploying on S3, we need to make some configurations on AWS S3 by creating a bucket and telling AWS that we are going to serve a static website. In the AWS console menu, choose the S3 service and create a bucket. Follow these steps to deploy a React application on AWS using the S3 service:

1. You will need to enter some configurations such as the **Bucket name** value and others, as shown in the following figure:

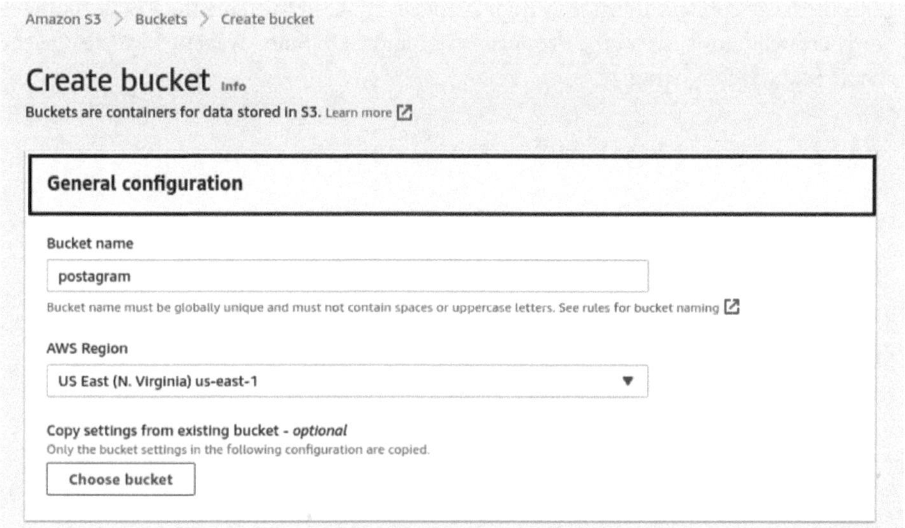

Figure 15.3 – General configuration for AWS S3 bucket

2. After that, you need to disable the **Block all public access** settings so that the React application is visible to the public:

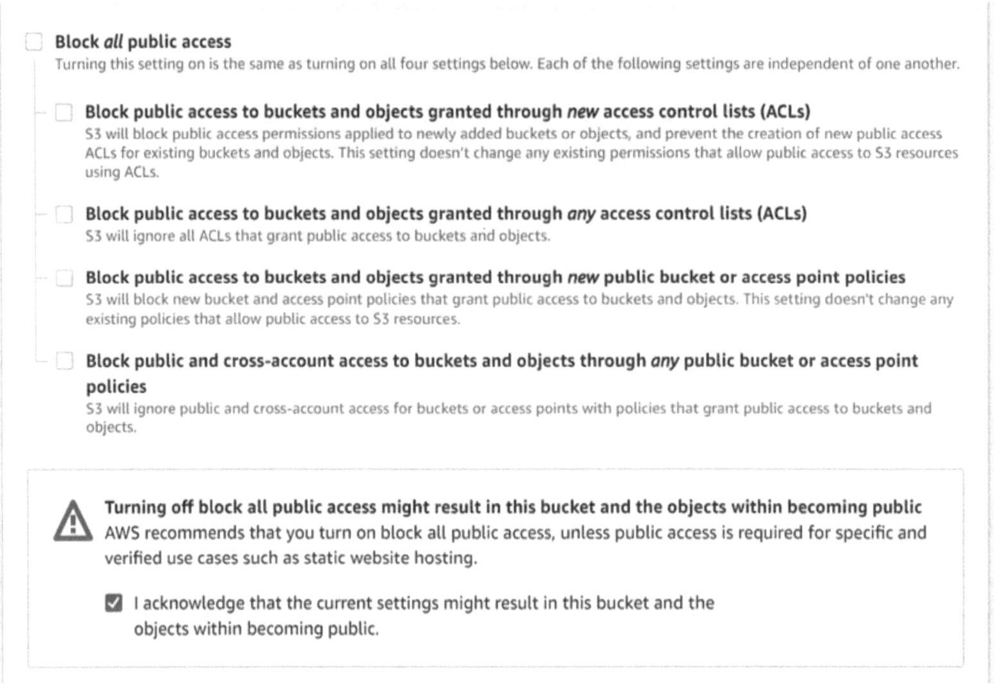

Figure 15.4 – Public access configuration

3. With the basic configurations now done, you can proceed to create the S3 bucket. Access the newly created bucket, select the **Properties** tab, and go to **Static website hosting**. On the page, enable **Static web hosting**:

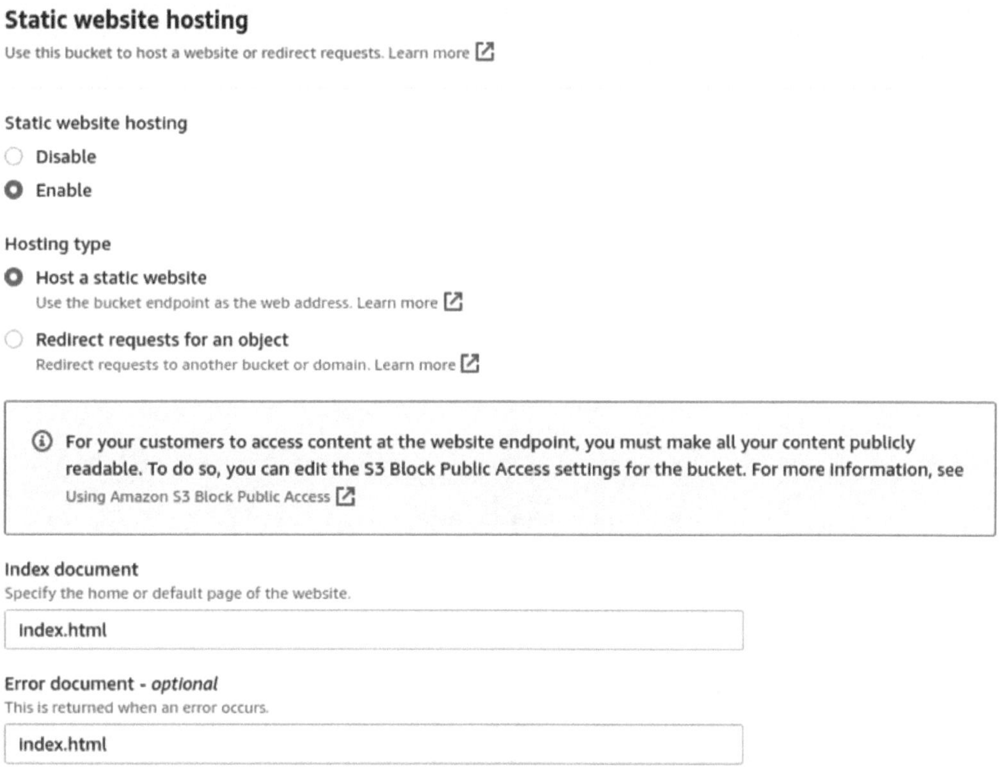

Figure 15.5 – Static website hosting configuration

4. You should also fill in the **Index document** and **Error document** fields. This will help with routing in the React application. Save the change, and you will see the bucket website endpoint, which will be the URL of your website:

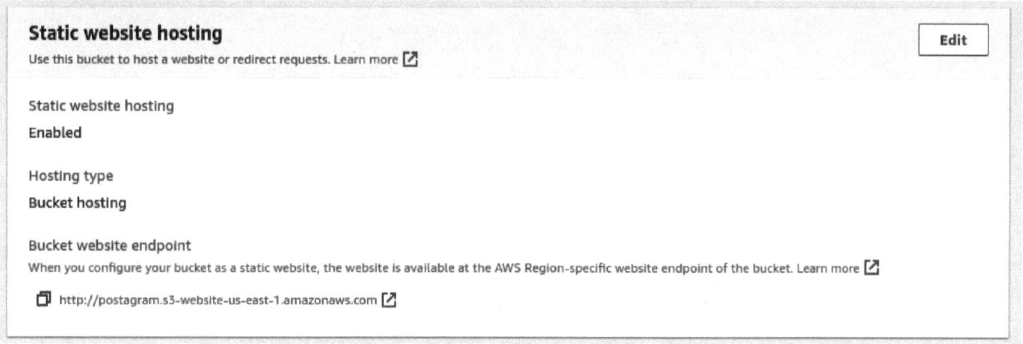

Figure 15.6 – Static website hosting configuration done

5. Finally, select the **Permissions** tab and select **Bucket Policy**. We will add a policy to grant public access to the bucket, like so:

```
{
      "Version": "2012-10-17",
      "Statement": [
            {
                  "Sid": "Statement1",
                  "Effect": "Allow",
                  "Principal": {
                        "AWS": "*"
                  },
                  "Action": "s3:GetObject",
                  "Resource": "arn:aws:s3:::postagram/*"
            }
      ]
}
```

   In your case, replace Postagram with the name of your React application.

6. Save the changes. You will notice that a piece of new information will appear next to the name of the bucket:

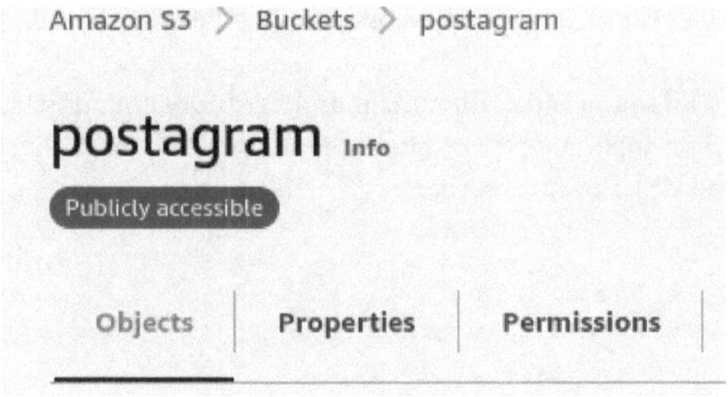

Figure 15.7 – Publicly accessible badge

7. Now, click on the **Upload** button and upload all content in the `build` directory of the React application. After the upload is finished, you will have a similar result to this:

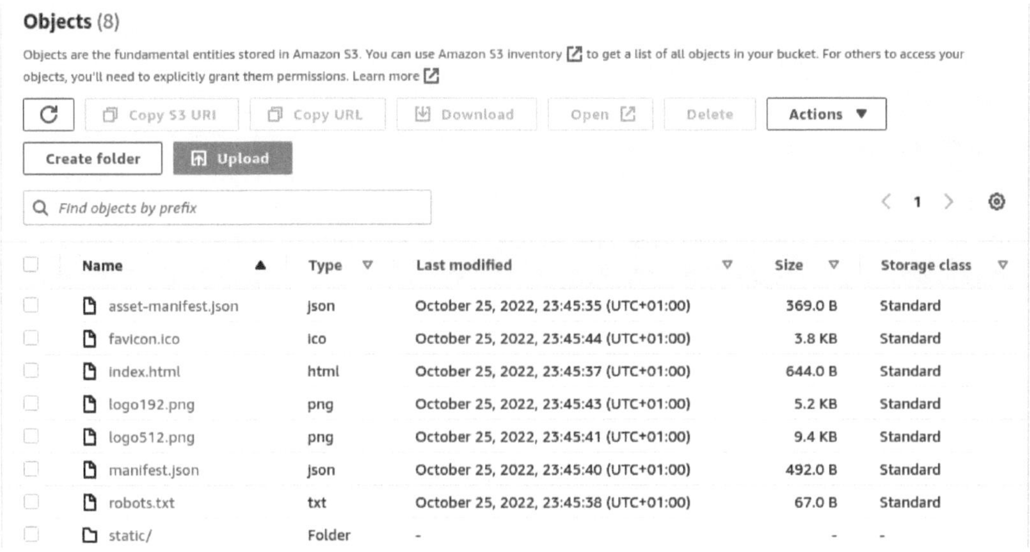

Figure 15.8 – Bucket content

8. Click on the bucket website endpoint, and you will access the Postagram React application in your browser:

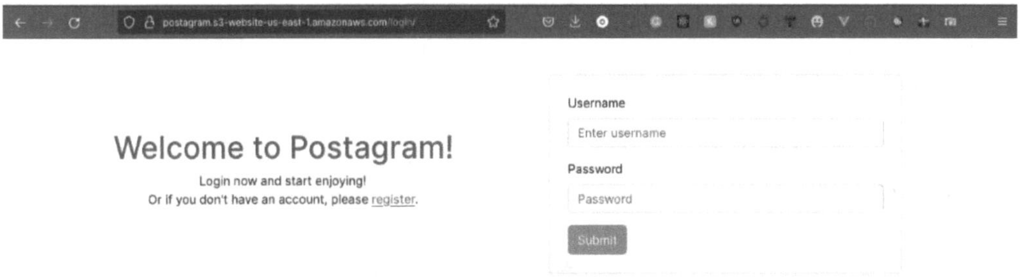

Figure 15.9 – Deployed React application

Great! We have deployed a React application on AWS using the S3 service. You will surely encounter **cross-origin resource sharing** (**CORS**) errors when trying to make some requests to the API. To resolve this issue, add the domain name of the link provided by AWS S3 for your static website to the `CORS_ALLOW_ORIGINS` environment variables in the `.env` file of the Django application on the AWS EC2 instance. The following is an example of how you can define the environment variable:

.env

```
CORS_ALLOW_ORIGINS="S3_WEBSITE_URL"
```

Then, in the `settings.py` file of the Django project, replace the line where you define `CORS_ALLOW_ORIGINS` with the following:

CoreRoot/settings.py

```
...
CORS_ALLOWED_ORIGINS = os.getenv("CORS_ALLOWED_ORIGINS", "").
split(",")
...
```

We have learned how to configure a bucket, change the policies for public access, and activate the website hosting feature of AWS S3. However, the deployment was done manually and, in the future, if you are pushing regularly, it might be a hassle to upload the change manually every time. In the next section, we will explore how to automate the deployment of a React application using GitHub Actions.

Automating deployment with GitHub Actions

In the previous chapter, we explored how GitHub Actions make the flow of deployment easier, more secure, and more reliable for developers. That is why in this chapter, we are also using GitHub Actions to automate the deployment of the React application.

There is a GitHub action for AWS called `configure-aws-credentials`. We will use this action to configure AWS credentials in the workflow to execute a command to upload the content of the `build` folder in the S3 bucket created earlier. But before that, we will follow the same workflow of CI/CD:

1. Install the dependencies of the project.
2. Run tests to make sure the application won't break in production and to ensure there are no regressions.
3. Run the `build` command to have a production-ready application.
4. Deploy on AWS S3.

Let's add a new workflow file in the repository for the deployment of the React application.

> **Important note**
>
> For this book, the Django application and the React application are in the same repository. The choice was made to make it easier for you to go through the code and the project. Thus, you will find two workflows in the `.github/workflows` directory. If you have split the code of the Django application and the React project into different repositories, make sure to not mix the GitHub Actions files.

Writing the workflow file

Inside the `.github/workflows` directory, create a file called `deploy-frontend.yml`. The first step, as usual, when writing a GitHub Actions file is to define the name of the workflow and the condition that will trigger this workflow:

.github/workflows/deploy-frontend.yml

```
name: Build and deploy frontend

on:
 push:
   branches: [ main ]
```

Let's then create a job called `build-test-deploy`. Inside this job, we will write the commands to install the React dependencies, run the tests, build the project, and deploy the application to S3. Let's start by injecting the environment variables:

.github/workflows/deploy-frontend.yml

```
jobs:
 build-test-deploy:
   name: Tests
   runs-on: ubuntu-latest
   defaults:
     run:
       working-directory: ./social-media-react
   steps:
     - uses: actions/checkout@v2

     - name: Injecting environment variables
       run: echo "REACT_APP_API_URL=${{ secrets.API_URL }}"
            >> .env
```

We can now add the commands to install the dependencies, run the tests, and build the application:

.github/workflows/deploy-frontend.yml

```
      - name: Installing dependencies
        run: yarn install

      - name: Running tests
        run: yarn test

      - name: Building project
        run: yarn build
```

And we can add the AWS credentials action to configure the AWS credentials in the workflow and run the command to deploy to S3:

.github/workflows/deploy-frontend.yml

```
      - name: Configure AWS Credentials
        uses: aws-actions/configure-aws-credentials@v1
        with:
          aws-access-key-id: ${{ secrets.AWS_ACCESS_KEY_ID
                             }}
          aws-secret-access-key:
            ${{ secrets.AWS_SECRET_ACCESS_KEY }}
          aws-region: us-west-1

      - name: Deploy to S3 bucket
        run: aws s3 sync ./build/ s3://postagram --delete
```

In the last command, we are uploading the content of the `build` directory to the Postagram bucket. While using this configuration, ensure to use the name of your S3 bucket. The GitHub actions file is written and can be deployed. Commit the changes and push them to the GitHub repository.

Congratulations! You have deployed a React application to AWS S3 using GitHub Actions.

We have successfully deployed the full-stack application we have been building in this book. We have deployed the Django API application on an AWS instance, deployed the React frontend on AWS S3, and automated CI/CD pipelines using GitHub Actions. However, before going fully live, we need to make some optimization on the backend and the frontend, secure the deployed version of the applications on AWS using HTTPS, and talk more about caching and SQL query optimization.

Summary

In this chapter, we have deployed the frontend React application on AWS. We have explored the AWS S3 service created and developed by AWS for storing objects on the internet. We have learned how to add environment variables to a React application but also how to have a production-ready bundle by building the application.

The production bundle has been used for deployment on AWS S3 using a bucket and configuring the bucket for static website hosting. And to make the deployment process smooth, we have created a GitHub action to automate the CI/CD pipeline for the React frontend project from building and testing to deploying the application on AWS S3.

In the next chapter, we will focus on the optimization of the Django API and the React frontend by optimizing queries, adding caching, adding a logout endpoint, and securing the communication between servers and the client using HTTPS.

Questions

1. What is AWS S3?
2. How to create an IAM user on AWS?
3. What is the command used to build a React application?
4. Where are the environment variables in a Node.js project retrieved from?

16

Performance, Optimization, and Security

In the previous chapters of the book, we have created a full stack application from scratch, starting with building and creating a REST API using Django and Django REST Framework and then creating a web interface with React to communicate with the API we created. We have also deployed applications on services such as AWS EC2 and AWS S3. However, we need to further investigate some important aspects of having an application deployed on the internet, such as performance checks, query optimization, frontend optimization, and finally, security aspects.

In this chapter, we will learn how to create a performant API with fewer SQL queries and faster API responses, how to serve the API and the React frontend over HTTPS using AWS CloudFront, and how to log out users using the API. In this chapter, we will cover the following points:

- Revoking JWT tokens
- Adding caching
- Optimizing the deployment of a React application
- Securing deployed applications with HTTPS with AWS CloudFront

Technical requirements

For this chapter, you need to have an active AWS account with access to services such as S3, EC2, and CloudFront. You can also find the code for this chapter at: `https://github.com/PacktPublishing/Full-stack-Django-and-React/tree/chap16`.

Revoking JWT tokens

In this book, we have implemented an authentication system using **JSON Web Tokens (JWTs)**, and because it is a stateless authentication system, most of the authentication flow is handled by the frontend. If we want to log the user out of the Postagram React application, we must clear the tokens from the local storage of the browser, and the user is automatically redirected to the login page. But even if the tokens are deleted from the browser, they are still active.

The refresh tokens have a longer life period, so if a hacker gets their hands on a refresh token, they can still request access tokens and make HTTP requests using someone else's identity. To avoid that, we will add a logout feature to invalidate access and refresh tokens from the server side.

The package used to add JWT authentication on the Django REST API (`djangorestframework-simplejwt`) supports blacklisting tokens, and that is the perfect feature we need here. Let's set up the required configurations for the logout feature, and let's add the feature to the Django REST API.

Adding a logout endpoint

In this section, we will write some code on the Django application to add an endpoint for logout:

1. In the `settings.py` file of the project, add the following entry to the `INSTALLED_APPS` list:

CoreRoot/settings.py

```
...
"corsheaders",
"rest_framework_simplejwt.token_blacklist",
...
```

2. After that, create a file called `logout.py` in the `core/auth/viewsets` directory. This file will contain the code for the `viewsets` and the logic to blacklist a token.
3. In this file, add the required imports and define the `LogoutViewSet` class:

core/auth/viewsets/logout.py

```
from rest_framework_simplejwt.tokens import RefreshToken,
TokenError
from rest_framework import viewsets, status, permissions
from rest_framework.exceptions import ValidationError
from rest_framework.response import Response


class LogoutViewSet(viewsets.ViewSet):
    authentication_classes = ()
    permission_classes = (permissions.IsAuthenticated,)
    http_method_names = ["post"]
```

The logout endpoint will only accept POST requests, as the client will be required to pass a refresh token within the body of the POST request. We also specify that only authenticated users have permission to access this endpoint.

4. Let's write the `create` method of the `LogoutViewSet` class:

core/auth/viewsets/logout.py

```
...
class LogoutViewSet(viewsets.ViewSet):
...
    def create(self, request, *args, **kwargs):
        refresh = request.data.get("refresh")
        if refresh is None:
            raise ValidationError({"detail":
              "A refresh token is required."})


        try:
            token = RefreshToken(request.data.get(
              "refresh"))
            token.blacklist()
            return Response(
              status=status.HTTP_204_NO_CONTENT)
        except TokenError:
            raise ValidationError({"detail":
              "The refresh token is invalid."})
```

In the preceding code, we ensure that the refresh token is present in the body of the request. Otherwise, we raise an error. Once the verification is done, we encapsulate the blacklisting logic in a `try/except` block:

- If the token is valid, then the token is blacklisted, and we return a response with a `204` `HTTP` status code.
- If there is an error related to the token, then the token is invalid, and we return a validation error.

5. Let's not forget to add the newly created `ViewSet` in the `routers.py` file and register a new route:

core/routers.py

```
...
from core.auth.viewsets import (
    RegisterViewSet,
    LoginViewSet,
    RefreshViewSet,
    LogoutViewSet,
)
...
router.register(r"auth/logout", LogoutViewSet,
basename="auth-logout")
```

6. Great! To follow best practices for building software, we must add a test for the newly added route in the `core/auth/tests.py` file:

core/auth/tests.py

```
...
def test_logout(self, client, user):
    data = {"username": user.username,
            "password": "test_password"}
    response = client.post(self.endpoint + "login/",
                           data)

    assert response.status_code == status.HTTP_200_OK

    client.force_authenticate(user=user)

    data_refresh = {"refresh":
      response.data["refresh"]}

    response = client.post(self.endpoint + "logout/",
      data_refresh)
    assert response.status_code ==
      status.HTTP_204_NO_CONTENT
```

In the preceding code, we log in to retrieve a refresh token and force the authentication for the user so we can access the logout endpoint. After that, we ensure that we have returned the right status code when the logout is successful.

7. Run the tests using the `pytest` command. If you are using Docker, then you can run the tests using this command:

```
docker-compose exec -T api pytest
```

With the logout endpoint ready, we can now make some modifications to the authentication logic (mostly the logout logic) in the React application.

Handling the logout with React

We have already handled the logout on the React application to a certain extent by just deleting the tokens from the local storage. There is nothing big to modify here, we will just add a function to make a request to the API, and if this request is successful, we will delete the tokens and the user from the local storage of the browser. The current logout logic on the React application is handled in the `NavigationBar` component:

src/components/NavBar.jsx

```
...
            <NavDropdown.Item
              onClick={userActions.logout}>
              Logout
            </NavDropdown.Item>
...
```

Inside the `useActions` Hook function, let's tweak the `logout` method to make an API call before deleting the user:

src/hooks/user.actions.js

```
...
  // Logout the user
  function logout() {
    return axiosService
      .post(`${baseURL}/auth/logout/`,
            { refresh: getRefreshToken() })
      .then(() => {
```

```
        localStorage.removeItem("auth");
        navigate("/login");
      });
  }
```

Once it is done, let's create a function in the `NavigationBar` component to handle the cases when there is an error from the API. We will display a toast HTML bloc on the page with the error message:

src/components/NavBar.jsx

```
import React, { useContext } from "react";
import { Context } from "./Layout";
...

function NavigationBar() {
  const { setToaster } = useContext(Context);

  const userActions = useUserActions();

  const user = getUser();

  const handleLogout = () => {
    userActions.logout().catch((e) =>
      setToaster({
        type: "danger",
        message: "Logout failed",
        show: true,
        title: e.data?.detail | "An error occurred.",
      })
    );
  };
...
```

Great! Our full stack application now supports logout. In the next section, we will discuss a recurrent topic when deploying a project online, caching.

Adding caching

In software computing, caching is the process of storing copies of files in a cache so they can be accessed more quickly. A **cache** is a temporary storage location that stores data, files, and information concerning software that is regularly requested.

A great example and explanation of caching comes from Peter Chester, who asked the audience at one of his speeches: *"What's 3,485,250 divided by 23,235?" Everyone fell silent for a moment, but someone pulled a calculator and yelled out the answer "150!"*. Then, Peter Chester asked the same question again, and this time, everyone was able to answer the question immediately.

This is a great demo of the concept of caching: *The computation is only done once by the machine and then saved in quick memory for faster access.*

It is a concept used widely by companies and primarly social media websites where millions of users access the same posts, videos, and files. It would be very primitive to hit the database whenever millions of people want to access the same information. For example, if a tweet is gaining traction on Twitter, it is automatically moved to cache storage for quick access. And, if you have an influencer such as Kim Kardashian posting a picture on Instagram, you should expect a lot of requests for this picture. Thus caching can be useful here to avoid thousands of queries on the database.

To recapitulate, caching brings the following benefits:

- Reduced load time
- Reduced bandwidth usage
- Reduced SQL queries on databases
- Reduced downtime

Now that we have an idea about caching and its benefits, we can implement the concept using Django and even Docker. But before that, let's quickly discuss the complexity caching brings to your application.

The cons of caching

You already know the advantages of using caching, mostly if your application is scaling or you want to improve load time and reduce costs. However, caching introduces some complexity to your system (it can also depend on the type of application you are developing). If your application is based on news or feeds, you might be in trouble, as you will need to define a good architecture for caching.

On the one hand, you have the chance to reduce load times by showing your users the same content for a period, but at the same time, your users might miss updates and maybe some important updates. Here, cache invalidation comes to the rescue.

Cache invalidation is the process of declaring cached content as invalid or stale. The content is invalidated, as it is no longer marked as being the most up-to-date version of a file. There are some methods available to invalidate a cache, as follows:

- **Purge (flush)**: Cache purging instantly removes the content from the cache. When the content is requested again, it is stored in the memory cache before returning it to the client.
- **Refresh**: A cache refresh consists of refreshing the same content from the server and replacing the content stored in the cache with the new version fetched from the server. This is done in the React application using **state-while-revalidate** (**SWR**). Each time a post is created, we call a refresh function to fetch data again from the server.
- **Ban**: A cache ban does not remove content from the cache immediately. Rather, the content is marked as blacklisted. Then, when the client makes a request, it is matched with the blacklist content, and if a match is found, new content is fetched again and updated in the memory cache before returning to the client.

With the cons of caching and how to invalidate the cache understood, you are well equipped to add caching to the Django application. In the next section, let's add caching to the Django API of Postagram.

Adding caching to the Django API

In the previous paragraphs, we have explored caching, its advantages, and the cons of the concept. Now, it's time to implement caching within our Django application. Django provides useful support for caching, which makes the configuration of caching within Django straightforward. Let's start by making the required configurations depending on your environment.

Configuring Django for caching

Using caching within Django requires configuring a memory cache. For the quickest read and write access, it is better to use a different data storage solution from SQL databases as SQL databases are known to be slower than memory databases (again, it depends on your needs). In this book, we will use Redis. Redis is an open source, in-memory data store used as a database, cache, streaming engine, and message broker.

We'll review the configurations you need to make to start using Redis in your Django project, whether you are using Docker or not. However, for the deployment, we'll use Docker for configuring Redis.

So, if you are not going to use Docker, you can install Redis using the following link: `https://redis.io/download/`.

> **Important note**
>
> If you are working in a Linux environment, you can check whether the service is running using the `sudo service redis-server status` command. If the service is not active, use the `sudo service redis-server start` command to start the Redis server. If you are using Windows, you will need to install or enable WSL2. You can read more at: `https://redis.io/docs/getting-started/installation/install-redis-on-windows/`.

After the installation on your machine, you can configure caching in Django using the `CACHES` setting in the `settings.py` file of the Django project:

CoreRoot/settings.py

```
...
CACHES = {
    'default': {
        'BACKEND': 'django_redis.cache.RedisCache',
        'LOCATION': 'redis://127.0.0.1:6379/1',
        'OPTIONS': {
            'CLIENT_CLASS':
              'django_redis.client.DefaultClient',
        }
    }
}
```

This configuration will require the installation of a Python package called `django-redis`. Install it by running the following command:

```
pip install django-redis
```

If you are working with Docker, you just need to add the following configurations:

1. Add the `django-redis` package to the `requirements.txt` file:

requirements.txt

```
django-redis==5.2.0
```

2. Add the `docker-compose.yaml` configuration. We will add a new image in the Docker configuration to make sure that the Django application requires `redis-server` to be ready before the API service starts running:

docker-compose.yaml

```
services:
  redis:
    image: redis:alpine
…
api:
...
  depends_on:
    - db
    - redis
...
```

3. Great! Add the following custom backend in the `settings.py` file of the Django project:

CoreRoot/settings.py

```
CACHES = {
    "default": {
        "BACKEND": "django_redis.cache.RedisCache",
        "LOCATION": "redis://redis:6379",
        "OPTIONS": {
            "CLIENT_CLASS":
              "django_redis.client.DefaultClient",
        },
    }
}
```

You will notice here that we are using `redis` as the host instead of `127.0.0.1`. This is because, with Docker, you can use the name of the service as a host. This is a better solution; otherwise, you will have to configure a static IP address for the services.

> **Important note**
>
> If you want to learn more about assigning a static IP address to your containers with Docker, you can read the following resource: `https://www.howtogeek.com/devops/how-to-assign-a-static-ip-to-a-docker-container/`.

Great! Now that we have configured Django for caching, let's build the caching system for the Postagram application.

Using caching on the endpoints

Caching depends a lot on the business requirements for how much time you want to cache the data. Well, Django provides many levels for caching:

- **Per-site cache**: This enables you to cache your entire website.
- **Template fragment cache**: This enables you to cache some components of the website. For example, you can decide to only cache the footer.
- **Per-view cache**: This enables you to cache the output of individual views.
- **Low-level cache**: Django provides an API you can use for interacting directly with the cache. It is useful if you want to produce a certain behavior based on a set of actions. For example, in this book, if a post is updated or deleted, we will update the cache.

Now that we have a better idea about the levels of caching Django provides, let's define the caching requirements for the Postagram API.

Our requirement is if there is a delete or an update on a comment or a post, the cache is updated. Otherwise, we return the same information in the cache to the user.

This can be achieved in many ways. We can use Django signals or directly add custom methods to the manager of the model's `Post` and `Comment` classes. Let's go with the latter. We will surcharge the `save` and `delete` methods of the `AbstractModel` class, so if there is an update on a `Post` or `Comment` object, we update the cache.

Inside the `core/abstract/models.py` file, add the following method on top of the file after the imports:

core/abstract/models.py

```
from django.core.cache import cache
...
def _delete_cached_objects(app_label):
    if app_label == "core_post":
        cache.delete("post_objects")
    elif app_label == "core_comment":
        cache.delete("comment_objects")
    else:
        raise NotImplementedError
```

The function in the preceding code takes an application label, and according to the value of this `app_label`, we invalidate the corresponding cache. For the moment, we only support caching for posts and comments. Notice how the name of the function is prefixed with a _. It is a coding convention to specify that this method is private and should not be used outside the file where it is declared.

Inside the `AbstractModel` class, we can surcharge the `save` method. Before the `save` method is executed, we invalidate the cache. It means that on operations such as `create` and `update`, the cache will be reset:

core/abstract/models.py

```
class AbstractModel(models.Model):
...


    def save(
        self, force_insert=False, force_update=False,
        using=None, update_fields=None
    ):
        app_label = self._meta.app_label
        if app_label in ["core_post", "core_comment"]:
            _delete_cached_objects(app_label)
        return super(AbstractModel, self).save(
            force_insert=force_insert,
            force_update=force_update,
            using=using,
            update_fields=update_fields,
        )
```

In the preceding code, we retrieve `app_label` from the `_meta` attribute on the model. If it corresponds to either `core_post` or `core_comment`, we invalidate the cache, and the rest of the instructions can proceed. Let's do the same for the `delete` method:

Core/abstract/models.py

```
class AbstractModel(models.Model):
...


    def delete(self, using=None, keep_parents=False):
```

```
        app_label = self._meta.app_label
        if app_label i" ["core_p"st", "core_comm"nt"]:
            _delete_cached_objects(app_label)
        return super(AbstractModel, self).delete(
          using=using, keep_parents=keep_parents)
```

Great. The cache invalidation logic has been implemented on the models. Let's add the logic for cache data retrieving on the viewsets of the `core_post` application and the `core_comment` application.

Retrieving data from the cache

The cache invalidation is ready, so we can freely retrieve data from the cache on the endpoints for the posts and the comments. Let's start with `PostViewSet` as the portion of code that will be written on `PostViewSet` and `CommentViewSet` will be the same. As a small exercise, you can write the logic for retrieving the cache for the comments.

Inside the `PostViewSet` class, we will rewrite the `list()` method. On the **Django REST framework** (**DRF**) open source repository, the code looks like this:

```
"""List a queryset"""
def list(self, request, *args, **kwargs):
    queryset = self.filter_queryset(self.get_queryset())


    page = self.paginate_queryset(queryset)
    if page is not None:
        serializer = self.get_serializer(page, many=True)
        return self.get_paginated_response(serializer.data)

    serializer = self.get_serializer(queryset, many=True)
    return Response(serializer.data)
```

In the preceding code, a `queryset` call is made to retrieve the data, and then this `queryset` call is paginated, serialized, and returned inside a `Response` object. Let's tweak the method a little bit:

core/post/viewsets.py

```
class PostViewSet(AbstractViewSet):
...

    def list(self, request, *args, **kwargs):
```

```
        post_objects = cache.get("post_objects")


        if post_objects is None:
            post_objects =
              self.filter_queryset(self.get_queryset())
            cache.set("post_objects", post_objects)

        page = self.paginate_queryset(post_objects)
        if page is not None:
            serializer = self.get_serializer(page,
                                             many=True)
            return self.get_paginated_response(
              serializer.data)


        serializer = self.get_serializer(post_objects,
                                         many=True)
        return Response(serializer.data)
```

In the preceding code, instead of doing a lookup on the database directly, we check the cache. If `post_objects` is `None` when making a query to the database, save `queryset` in the cache and finally proceed to return the cache objects to the user.

As you can see, the process is very simple. You just need to have a robust caching strategy. You can do the same for `CommentViewSet` as an exercise. You can check the code at this link to compare your results: `https://github.com/PacktPublishing/Full-stack-Django-and-React/blob/chap16/core/comment/viewsets.py`.

In this section, we have explored the benefits of caching, and we have implemented caching in the Django application. In the next section, we will see how to optimize the React build using tools such as `webpack`.

Optimizing the React application build

In the previous chapter, we successfully built the React application and made the deployment on AWS S3. However, we could have done better in terms of optimization and performance. In this section, we will use the famous webpack module builder to optimize the React build of Postagram.

There are a lot of advantages of using webpack in React:

- **It speeds up development and build times**: Using webpack in development enhances the speed of fast reload of React.
- **It provides minification**: Webpack automatically minimizes the code without changing the functionalities. This results in a faster load on the browser side.
- **Code splitting**: Webpack converts JavaScript files into modules.
- **It eliminates dead assets**: Webpack only builds the images and CSS that your code uses and needs.

Let's start by integrating webpack into the project.

Integrating webpack

Follow these steps to integrate webpack into your project:

1. Inside the React project, run the following command to add the `webpack` and `webpack-cli` packages:

```
yarn add -D webpack webpack-cli
```

2. Once the installation is done, modify the `package.json` scripts:

package.json

```
...
"scripts": {
    "start": "react-scripts start",
    "build": "webpack --mode production",
    "test": "react-scripts test",
    "eject": "react-scripts eject"
  },
...
```

Also, we need to install Babel, which is a JavaScript compiler that converts next-generation JavaScript code into browser-compatible JavaScript.

3. In the React project, Babel will convert the React components, the ES6 variables, and JSX code to regular JavaScript so old browsers can render the components correctly:

```
yarn add -D @babel/core babel-loader @babel/preset-env @
babel/preset-react
```

`babel-loader` is the webpack loader for Babel, `babel/preset-env` compiles with JavaScript to ES5, and `babel/preset-react` is for compiling JSX to JS.

4. Then create a new file called `.babelrc`:

```
{
  "presets": ["@babel/preset-env",
              "@babel/preset-react"]
}
```

5. Then create a new file called `webpack.config.js`. This file will contain the configurations for webpack. Before writing the configuration, add some plugins for optimizing HTML, CSS, and copy files:

```
yarn add -D html-webpack-plugin html-loader copy-webpack-
plugin
```

6. And then add the following configuration on `webpack.config.js`:

webpack.config.js

```
const path = require("path");
const HtmlWebPackPlugin = require("html-webpack-plugin");
const CopyPlugin = require("copy-webpack-plugin");
const webpack = require("webpack");

module.exports = {
 module: {
   rules: [
     {
       test: /\.(js|jsx)$/,
       exclude: /node_modules/,
       use: {
         loader: "babel-loader",
       },
     },
     {
       test: /\.css$/i,
       use: ["style-loader", "css-loader"],
     },
   ],
 },
};
```

The preceding code above tells webpack to send all files in `.js` and `.jsx` through `babel-loader`.

7. Let's add another configuration called `resolve` to generate all the possible paths to the module. For example, webpack would then proceed to look up each of those paths until it finds a file:

webpack.config.js

```
...
  resolve: {
    modules: [path.resolve(__dirname, "src"),
              "node_modules"],
    extensions: ["", ".js", ".jsx"],
  },
};
```

8. Let's add the configuration for the plugins we will use in this project:

webpack.config.js

```
...
  plugins: [
    new HtmlWebPackPlugin({
      template: "./public/index.html",
      filename: "./index.html",
    }),
    new CopyPlugin({
      patterns: [
        {
          from: "public",
          globOptions: {
            ignore: ["**/*.html"],
          },
        },
      ],
    }),
    new webpack.DefinePlugin({ process: {env: {}} }),
  ],
  output: {
```

```
    publicPath: '.',
  },
};
```

In the preceding code, we have added plugin configurations for the following:

- `html-loader`: This will send the HTML files through `html-loader`
- `copy`: This will copy the content of the public file to the `dist` file
- `define`: This plugin declares the `process` object so we can access environment variables in the production environment

9. Once it is done, run the `build` command:

```
yarn build
```

Webpack will take control and build the React application in the `dist` directory:

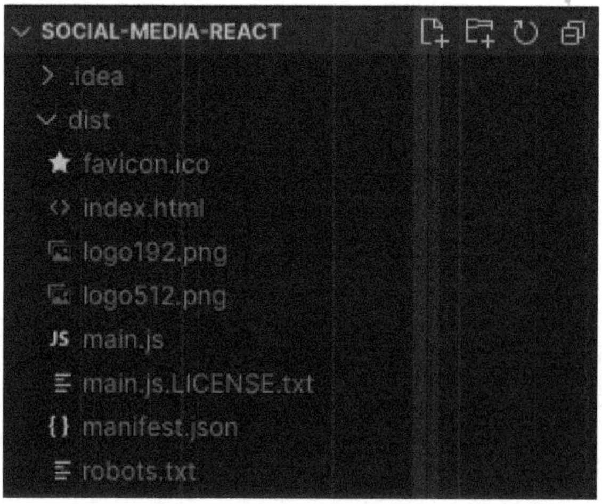

Figure 16.1 – The content of the dist directory

Great! You can push the changes made to GitHub, and the code will be deployed on AWS S3. To make the testing and build faster, we will change the package manager from `yarn` to `pnpm`. The next section is optional, but it will help you with a faster build for your React application.

Using pnpm

`pnpm` is a replacement for the `npm` JavaScript package manager, which is built on top of `npm`, and is much faster and more efficient. It provides advantages such as disk space efficiency, improved speed, and better security. The `pnpm` package manager is the one to use if you want to spend less time building and making cuts to the minutes spent on the GitHub Actions.

Let's install pnpm on our machine:

```
npm install -g pnpm
```

After that, we can generate a `pnpm-lock.yaml` file. We can generate this file from another manager's lock file, in our case, from the `yarn.lock` file:

```
pnpm import
```

Figure 16.2 – Result of pnpm import

A new file will be generated in the directory of the React project. Then, modify the `deploy-frontend.yml` file to configure for pnpm usage:

.github/workflows/deploy-frontend.yml

```
jobs:
 test:
   name: Tests
   runs-on: ubuntu-latest
   defaults:
     run:
       working-directory: ./social-media-react
   steps:
     - uses: actions/checkout@v3
     - uses: pnpm/action-setup@v2.2.4
       with:
         version: 7
     - name: Use Node.js 16
       uses: actions/setup-node@v3
       with:
         node-version: 16
```

```
        cache: 'pnpm'
        cache-dependency-path:
          ./social-media-react/pnpm-lock.yaml
```

After that, just replace `yarn` with `pnpm` in the `deploy-frontend.yml` file. You will notice a faster build for the React application.

In this section, we have covered pnpm and webpack and how they can boost the performance of the React application. In the next section, we will learn how to secure HTTP requests using AWS CloudFront.

Securing deployed applications with HTTPS with AWS CloudFront

When we have deployed the backend and the frontend on AWS S3, the applications are served through HTTP. Basically, our full stack application is not secured on the internet, and we are vulnerable. According to the **Open Web Application Security Project** (**OSWAP**) description of Insecure Transport (`https://owasp.org/www-community/vulnerabilities/Insecure_Transport`), our application is vulnerable to the following attacks:

- Attacks targeting login credentials, session IDs, and other sensitive information
- Bypassing **Secure Sockets Layer** (**SSL**) protocol by entering HTTP instead of HTTPS at the beginning of the URL in the browser
- Sending non-protected URLs of authentication pages to users to trick them into authenticating via HTTP

AWS EC2 and AWS S3 don't serve content through HTTPS by default. But AWS also has a service called CloudFront that can help you serve your applications via HTTPS, plus it also makes the content available globally.

AWS CloudFront is a content delivery network service, and in the next section, we will configure the AWS S3 bucket hosting the React application with AWS Cloudfront.

Configuring the React project with CloudFront

Follow these steps to configure our React project with CloudFront:

1. On the AWS dashboard, select the **CloudFront** service in the AWS console and click on **Create Distribution**.

2. Copy the origin of your website hosted on AWS and paste it into the **Origin domain** name field:

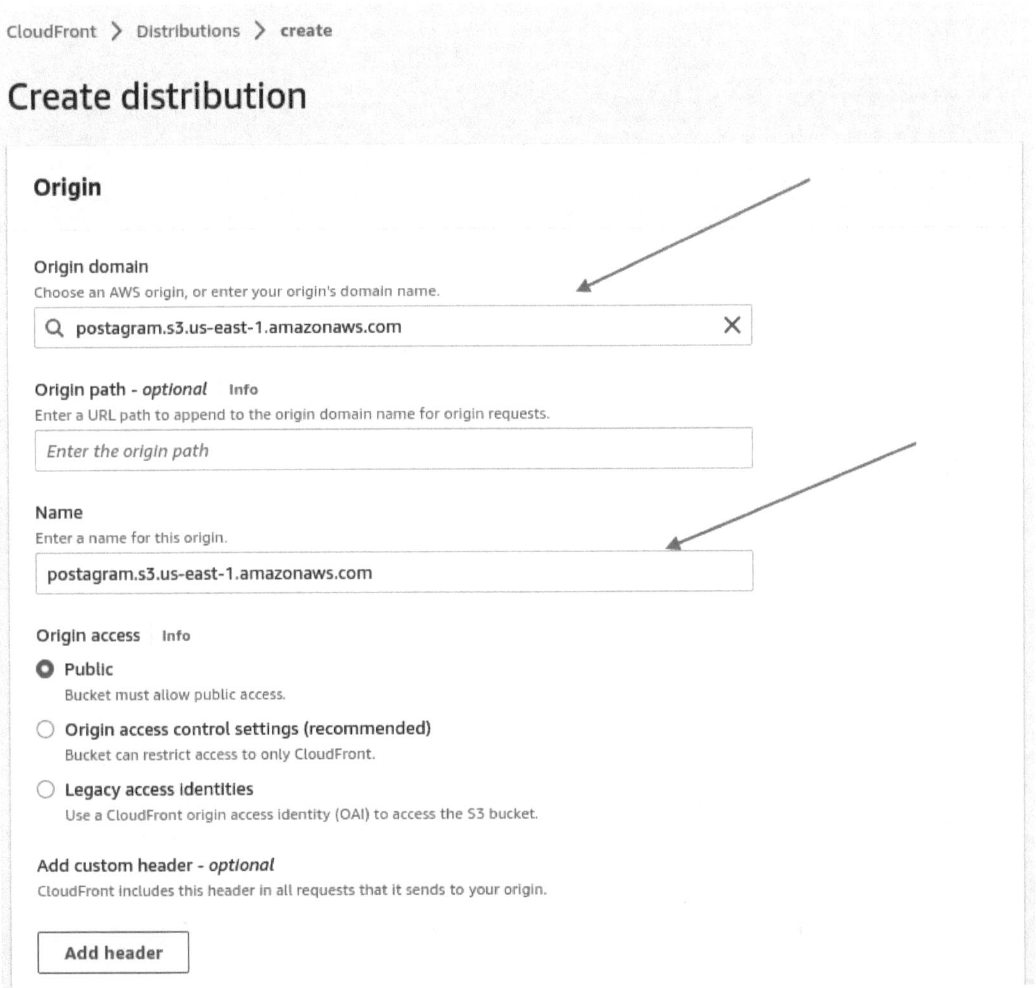

Figure 16.3 – Origin configuration of the CloudFront distribution

3. Next, configure the default cache behaviors:

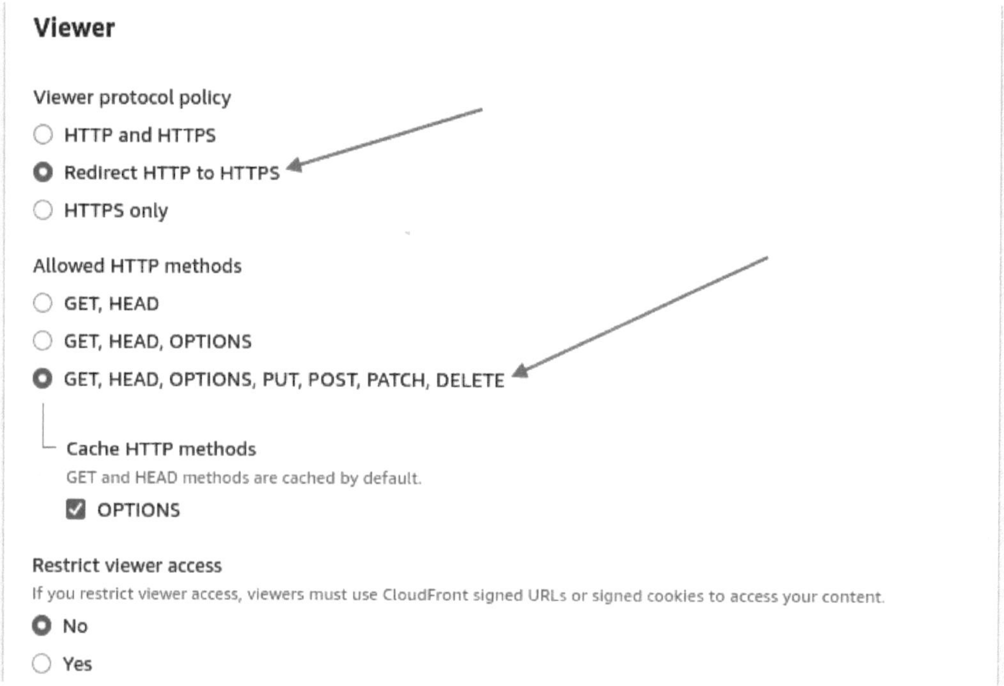

Figure 16.4 – Viewer configuration of the CloudFront distribution

4. Once the cache configuration is done, create the distribution. AWS will take some time to create the distribution and once it is done, click on the distribution **ID** field to copy the URL:

Figure 16.5 – List of the CloudFront distribution

5. Once **Status** changes to **Enabled**, click on the distribution **ID** field to access more details about the distribution and copy the distribution domain name:

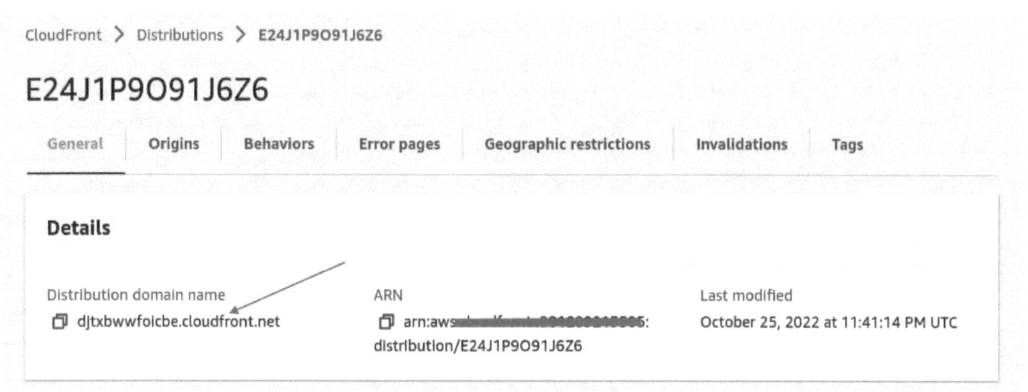

Figure 16.6 – Details about the created CloudFront distribution

The CloudFront distribution URL will return the React application over HTTPS. Great, the React application is secured on the internet and well distributed worldwide. Great! We have successfully secured our application over HTTPS using AWS CloudFront. From now, you can build a full stack application with Django and React, assure code quality with tests and linting, automate **continuous integration** and **continuous delivery** (**CI/CD**) pipelines using GitHub Actions and use AWS services such as S3, EC2, and CloudFront to deploy and serve your web application around the world.

Summary

In this chapter, we have covered some important points about optimizations and security. We have implemented a logout endpoint to blacklist tokens, added caching to the Django application using Redis, optimized the backend build using webpack, and secured the full stack application over HTTPS using AWS CloudFront. And that's the final touch of this book.

We have covered how to build a powerful and robust full stack application using Django and React. We have covered how to create a project from scratch, build an API secured with JWT tokens, build a frontend application with React and Bootstrap, and deploy the applications on AWS. We have explored Docker and tools such as GitHub Actions to make the development and deployment process secure, faster, and automated. You can now build and deploy a full stack application using Django and React!

We are now at the end of this book, and if you are looking for best practices and what to learn next, feel free to go through the *Appendix* directly after this chapter.

Questions

1. What is AWS CloudFront?
2. What are the cache invalidation strategies?
3. Why is logging important?

Appendix

Every successful application will eventually need to scale, and this process can cause resource issues and more optimization problems. In this appendix, I will list what you can read to deepen your studies after this book so you can become a better full stack developer.

Logging

Logging is the action of collecting information about an application as it performs different tasks or events. In the development process of an application, if you have a bug, you can use `print()` or `console.log()` to identify the issues. Even better, with `DEBUG` as `true` in Django, you have access to the whole traceback of a `500` error. Once your project deployed in production, this is no longer the case. You can implement logging in files using the default logging package provided by Python; Django has full support that you can explore in the official documentation at `https://docs.djangoproject.com/en/4.1/topics/logging/`. If you are looking to get real-time notifications when you have a `500` error, you can connect your backend to services such as Sentry, Datadog, or Bugsnag.

Database queries optimization

The Django ORM is a very flexible and powerful tool, and it can be used well or badly. Databases are important in your full stack applications and the fewer queries you make, the better it is for the high availability of the SQL database. Django provides many methods you can study and explore if you need to optimize database queries. You can read more at `https://docs.djangoproject.com/en/4.1/topics/db/optimization/`.

Security

If you are deploying a web application on the internet, it's important to ensure that you have a secure application. In the beginning, you don't really need a lot, but you do need to ensure that your system is secured against the top 10 threats listed by OWASP. You can learn more about this at the following link: `https://owasp.org/www-project-top-ten/`.

Answers

Chapter 1

1. A **Representational State Transfer** (**REST**) API is a web architecture and a set of constraints that provide simple interfaces to interact with resources, allowing clients to retrieve or manipulate them using standard HTTP requests.
2. Django is a Python web framework that enables the fast development of secure and maintainable websites. It follows the **Model-View-Controller** (**MVC**) architectural pattern and emphasizes reusability and pluggability.
3. To create a Django project, you need to have Django installed on your OS. Once you have it installed, you can use the following command to create a new Django project:

   ```
   django-admin startproject DjangoProject
   ```

   The preceding command will create a Django project with the name `DjangoProject`
4. Migrations are Django's way of synchronizing changes you make to your models (adding a field, deleting a model, etc.) into your database.
5. A virtual environment in Python is a tool to keep the dependencies required by different projects in separate places by creating isolated python virtual environments for them. This is useful in case of different projects and when you want to avoid conflicting dependencies.

Chapter 2

1. **JSON Web Token** (**JWT**) is a JSON object meant of representing claims to be transferred between two parties. JWT is often used to authenticate users in REST APIs.
2. **Django Rest Framework** (**DRF**) is a third-party package for Django that makes it easy to build, test, debug, and maintain RESTful APIs written using the Django framework.
3. A Django model is a Python class that represents a database table, and it defines the fields and behaviors of the data you're storing.
4. Serializers in DRF are used to convert complex data types, such as Django model instances or QuerySets, into JSON, XML, or other content types. Serializers also provide deserialization, which allows parsed data to be converted back into complex types.

5. Viewsets in DRF are classes that provide actions on model-backed resources. Viewsets are built on top of Django's class-based views and provide actions like `list`, `create`, `update`, and `delete`.
6. DRF routers provide a simple, quick, and consistent way of wiring viewsets to URLs. It allows you to automatically generate the URL conf for your API views.
7. A refresh token is a token that is issued by an authentication server and is used to obtain a new access token. Refresh tokens are used to keep the user authenticated indefinitely, by periodically obtaining a new access token.

Chapter 3

1. Some common database relationships in relational databases are:
   - **One-to-one**: This relationship is used when one record in a table is related to only one record in another table.
   - **One-to-many**: This relationship is used when one record in a table is related to multiple records in another table.
   - **Many-to-many**: This relationship is used when multiple records in one table are related to multiple records in another table.
2. Django REST permissions are used to control access to specific actions on specific viewsets. They can be used to restrict who can view, add, change, or delete data in your REST API.
3. In DRF, you can use the `LimitOffsetPagination` class to paginate the results of an API response. To use this class, you can include it in `REST_FRAMEWORK` in the `settings.py` file of your project.
4. To use the Django shell, you need to open the command line in the root directory of your Django project, and then run the following command:

```
python manage.py shell
```

Chapter 4

1. A nested route is a URL endpoint that represents a relationship between two or more resources. For example, in a social media application, you might have a route for all posts and another route for a specific post's comments. The comments route would be nested within the post route, allowing you to access the comments for a specific post.
2. `drf-nested-routers` is a package for DRF that allows you to easily create nested routes for your API. It automatically creates the appropriate URLs for related resources and allows you to nest your views within other views.

3. The `partial` attribute on the `ModelSerializer` can help you determine whether the user is submitting all the fields of the resource on an HTTP request for mutating like `PUT`, `PATCH`, or `DELETE`.

Chapter 5

1. Testing is a process of verifying that a system or software behaves in the way that it is expected to. Testing can be done manually or automatically.
2. A unit test is a test that verifies the functionality of a small and isolated piece of code, usually a single function or a method.
3. The testing pyramid is a concept that describes the balance between different types of tests in a software project. It suggests that most of the tests should be unit tests, which are fast and isolated, followed by a smaller number of integration tests, which test the interactions between different units of code, and a small number of end-to-end tests, which test the entire system.
4. Pytest is a popular testing framework for Python that makes it easy to write small, focused unit tests and provides many useful features such as test discovery, test parametrization, fixtures, and powerful and expressive assertion syntax.
5. A Pytest fixture is a way to provide data or set up resources that are needed for your tests. Fixtures are defined using the `@pytest.fixture` decorator and can be passed as arguments to test functions, allowing you to write more expressive and maintainable tests.

Chapter 6

1. Node.js is a JavaScript runtime built on Chrome's V8 JavaScript engine. It allows developers to run JavaScript on the server side to build fast and scalable network applications. Yarn is a package manager for Node.js, like npm but it is faster and more secure and provides a more consistent experience across different environments.
2. Frontend development is the process of building the user interface of a software application. In web development, it involves using languages such as **HTML**, **CSS**, and **JavaScript** to create the visual elements, layout, and functionality of a website.
3. To install Node.js, you can download the installer package from the official Node.js website (`https://nodejs.org/`) and then run it.
4. **Visual Studio Code** (**VS Code**) is a free, open-source code editor developed and maintained by Microsoft. It is a popular choice among developers for its support for multiple languages, debugging, and integrated Git control.
5. In VS Code, you can install extensions by clicking on the **Extensions** icon in the **Activity Bar** on the side of the editor, or by typing *Ctrl* + *Shift* + *X* (*Cmd* + *Shift* + *X* on macOS) to open the **Extensions** pane. You can then search for and install any installation you need.

6. Hot reloading is a feature that allows you to see the changes you make to your code immediately in the browser, without having to manually refresh the page. This makes development faster and more efficient, as you can see the effects of your changes in real-time.
7. To create a React application with `create-react-app`, you first need to have Node.js and yarn installed on your OS. Then, you can use `yarn` to create a new React application by running the following command in your terminal:

```
yarn create react-app my-app
```

Chapter 7

1. `localStorage` is an API provided by web browsers that allow developers to store **key-value** pairs of data locally on the client side. The data stored in `localStorage` persists even when the browser is closed, or the computer is restarted.
2. `React-Router` is a popular library for client-side routing in React. It allows you to declaratively map your application's component structure to specific URLs, making it easy to navigate between pages and manage the browser history
3. To configure a protected route in React, you can use React-Router's `<Route>` component along with a **higher-order component** (**HOC**) or a custom Hook that checks if the user is authenticated before rendering the protected component. For example:

```
function ProtectedRoute({ children }) {
  const user = getUser();

  return user ? <>{children}</> : <Navigate to=»/login/»
/>;
}
```

4. A React Hook is a special function that allows you to use state and other React features in a functional component. Hooks were introduced in React 16.8 to make it easier to write and manage stateful logic in functional components.
5. Some examples of React Hooks are:
   - `useState`: allows you to add a state to a functional component.
   - `useEffect`: allows you to run side effects such as fetching data or subscribing to an event in a functional component.
   - `useContext`: allows you to access the context values from a functional component.

6. The two rules of React Hooks are:

   - Only call Hooks at the top level. Don't call Hooks inside loops, conditions, or nested functions.
   - Only call Hooks from React function components. Don't call Hooks from regular JavaScript functions.

Chapter 8

1. A modal is a dialog box/pop-up window that is displayed on top of the current page. Modals are used to display content that requires the user's attention or input, such as forms, images, videos, or alerts.
2. In React, a **prop** (short for **property**) is a way to pass data from a parent component to a child component. Props are passed as attributes on a JSX element, and they can be accessed inside the child component using the `props` object.
3. The `children` element in React is a special prop that is used to pass content between elements. It is used to nest UI elements inside of other elements, and it can be accessed using the `props.children` property inside of the parent component.
4. A wireframe is a simplified visual representation of a web page or application, used to communicate the layout, structure, and functionality of a user interface.
5. The `map` method is an array method in JavaScript that is used to iterate over an array and create a new array with the results of a function applied to each element of the original array. It can also be used in JSX to map over an array and create a new set of elements.
6. The `mutate` method on SWR objects allows you to programmatically update the data in the cache, without waiting for the revalidation to happen. The `mutate` method triggers a re-render on the components that are using the data in the cache, updating the UI to reflect the new data.

Chapter 9

1. The `useParams` Hook is a built-in Hook in React Router that allows you to access the dynamic parameters passed in the URL of a route. It returns an object containing the **key-value** pairs of the parameters in the parameters.
2. In React, you can write a route that can support parameter passing by using the `:` syntax in the path of the route. For example, you can have `post/:postId` where `postId` is an URL parameter.
3. The `useContext` Hook is a built-in hook in React that allows you to access a context value within a functional component. This can be useful for sharing data across multiple components without having to pass props down through multiple levels of the component tree.

Chapter 10

1. The `FormData` object is a built-in JavaScript object that allows you to construct and send `multipart/form-data` requests. It can be used to upload files or other forms of binary data, as well as to send **key-value** pairs of data. The `FormData` object can be passed as the body of an `XMLHttpRequest` or `fetch` request, and it will automatically set the appropriate `Content-Type` header.
2. In Django, the `MEDIA_URL` setting is used to specify the URL at which user-uploaded media files will be served.
3. The `MEDIA_ROOT` setting in Django is used to specify the filesystem path where user-uploaded media files will be stored.

Chapter 11

1. The `render` method of the **React Testing Library** (**RTL**) is a utility function that allows you to render a component and its children as a tree of DOM nodes. The `render` method can be used to test the behavior and output of a component in a real-world-like environment.
2. Jest is a JavaScript testing framework that allows you to write and run unit tests for JavaScript code, including React components.
3. The `data-testid` attribute is a special attribute that allows you to add an identifier to an element for the purpose of testing. This attribute can be used to query the element in a test and make assertions about its state or behavior.
4. Some drawbacks of snapshot testing are:
   - Snapshots can become stale over time as the component changes, and they need to be updated manually.
   - Snapshot tests can be difficult to understand, as they often show the entire component tree, which can be large and complex.
5. To trigger user events in a React test suite, you can use React Testing Library `fireEvent` and `userEvent` methods.

Chapter 12

1. In Git, a branch is a separate line of development that allows one or multiple developers to work on different features or bug fixes simultaneously without interfering with each other's work. Branches are also used to isolate changes and make it easy to merge them back into the main codebase or branch.
2. Git is a **version control system** (**VCS**) that allows developers to track changes in their code over time, collaborate with others, and revert to previous versions if needed. GitHub is a web-based hosting service for Git repositories.

3. An HTTP Host header attack is a type of web application attack that exploits a vulnerability in the way some web servers handle the HTTP Host header. The HTTP Host header is used to specify the domain name of the website that the user is trying to access. By manipulating the Host header, an attacker can trick a vulnerable web server into serving content from a different domain, potentially exposing sensitive information, or allowing the attacker to perform actions on the user's behalf.
4. In Django, the `SECRET_KEY` setting is used to provide a secret key that is used to secure certain aspects of the Django framework, such as session management, password hashing, and the generation of cryptographic signatures. As it is a sensible piece of information, the value should be stored using environment variables.

Chapter 13

1. Docker is a platform for developing, shipping, and running applications that uses containerization technology to package an application and its dependencies into a single, portable container that can run on any platform that supports Docker. Containers provide a lightweight, isolated environment for running applications, which makes it easy to move them between development, staging, and production environments.
2. Docker Compose is a tool for defining and running multi-container Docker applications. It allows you to use a single `docker-compose.yml` file to configure and start multiple services (containers) that make up your application. This makes it easy to manage the dependencies and configuration of a complex application.
3. The main difference between Docker and Docker Compose is that Docker is a platform for creating, shipping, and running containers, while Docker Compose is a tool for defining and running multi-container applications. Additionally, Docker Compose relies on Docker to create and run the containers.
4. Virtualization is a technology that allows you to run multiple operating systems on a single physical machine, by creating virtual machines that emulate the hardware of a physical computer. Each virtual machine runs its operating system, and applications running inside the virtual machines are isolated from each other. Containerization is a technology that allows you to package an application and its dependencies into a single, portable container that can run on any platform. Containers are lightweight, isolated environments that share the host operating system kernel, which makes them faster and more efficient than virtual machines.
5. An environment variable is a value that can be passed to an operating system or application at runtime. It allows you to configure system-wide settings or to pass information to an application without hard-coding it in the source code. Environment variables can be used to set configuration options, such as the location of a file or the value of a secret key, and they can be easily changed without modifying the application's code.

Chapter 14

1. The differences between **Continuous Integration** (**CI**) and **Continuous Deployment** (**CD**) are:

   - CI is a software development practice in which developers integrate code into a shared repository multiple times a day. Each integration is verified by an automated build and test process to catch errors early.
   - CD is an extension of CI that goes a step further and automatically deploys the code changes to production after they pass the automated build and test process. The goal of CD is to make sure that the code is always in a releasable state and to shorten the time between code being written and it being available to end-users.

2. GitHub Actions is a feature provided by GitHub that allows developers to automate their software development workflows, such as building, testing, and deploying code. These workflows are defined in YAML files and can be triggered by various events such as a push to a branch, a pull request, or a scheduled time. Developers can use GitHub Actions to automate their CI/CD workflows.
3. CD is the practice of automatically building, testing, and deploying code changes to different environments after they pass the automated build and test process. It is an extension of CI, and the goal is to ensure that the code changes are always in a releasable state, so that they can be deployed to production at any time.

Chapter 15

1. **Amazon Simple Storage Service** (**S3**) is an object storage service provided by **Amazon Web Services** (**AWS**) that allows you to store and retrieve large amounts of data.
2. To create an **Identity and Access Management** (**IAM**) user on AWS, you can use the AWS Management Console. Here's an example of how to create an IAM user using the AWS Management Console:

   I. Log in to the AWS Management Console
   II. Open the IAM console.
   III. In the **Navigation** pane, choose **Users** and then choose **Add user**.
   IV. Type the username and select the **AWS access type**.
   V. Choose **Permissions**.
   VI. Choose **Add user to a group**, create group, or **Add existing groups** as appropriate.
   VII. Choose **Tags**.
   VIII. Choose **Review**.
   IX. Choose **Create user**.

3. The command used to build a React application is `react-scripts build`. This command will take all the code and assets in your application and create a production-ready build that can be deployed on a web server.
4. In a Node.js or more specifically a React project, environment variables are typically retrieved using the `process.env` object. For example, you can access the value of an environment variable named `VARIABLE` using `process.env.VARIABLE`.

Chapter 16

1. Amazon CloudFront is a **content delivery network** (**CDN**) provided by AWS. It allows you to distribute content, such as web pages, images, videos, and more, to users across the world by caching the content on servers located in various geographic locations. CloudFront can be used to deliver content from a variety of origins, such as an S3 bucket or a custom origin.
2. There are several strategies for cache invalidation in Django:
   - **Per-site cache**: This enables you to cache your entire website.
   - **Template fragment cache**: This enables you to cache some components of the website. For example, you can decide to only cache the footer.
   - **Per-view cache**: This enables you to cache the output of individual views.
   - **Low-level cache**: Django provides an API you can use for interacting directly with the cache. It is useful if you want to produce a certain behavior based on a set of actions.
3. Logging is important because it allows you to track the activity of your system, troubleshoot issues, and gather data for analysis. Logs provide a detailed history of what has happened in your system, including events such as user actions, system failures, and performance metrics. This information can be used to identify trends, detect patterns, and troubleshoot problems.

Index

A

B

C

D

E

F

G

H

I

J

K

L

M

N

O

P

R

S

T

Other Books You May Enjoy

If you enjoyed this book, you may be interested in these other books by Packt:

Becoming an Enterprise Django Developer

Michael Dinder

ISBN: 978-1-80107-363-9

- Use Django to develop enterprise-level apps to help scale your business
- Understand the steps and tools used to scale up a proof-of-concept project to production without going too deep into specific technologies
- Explore core Django components and how to use them in different ways to suit your app's needs
- Find out how Django allows you to build RESTful APIs
- Extract, parse, and migrate data from an old database system to a new system with Django and Python
- Write and run a test using the built-in testing tools in Django

Full Stack FastAPI, React, and MongoDB

Marko Aleksendrić

ISBN: 978-1-80323-182-2

- Discover the flexibility of the FARM stack
- Implement complete JWT authentication with FastAPI
- Explore the various Python drivers for MongoDB
- Discover the problems that React libraries solve
- Build simple and medium web applications with the FARM stack
- Dive into server-side rendering with Next.js
- Deploy your app with Heroku, Vercel, Ubuntu Server and Netlify
- Understand how to deploy and cache a FastAPI backend

Packt is searching for authors like you

If you're interested in becoming an author for Packt, please visit `authors.packtpub.com` and apply today. We have worked with thousands of developers and tech professionals, just like you, to help them share their insight with the global tech community. You can make a general application, apply for a specific hot topic that we are recruiting an author for, or submit your own idea.

Hi!

I am Kolawole Mangabo, author of *Full Stack Django and React*. I really hope you enjoyed reading this book and found it useful for increasing your productivity and efficiency in Django and React.

It would really help me (and other potential readers!) if you could leave a review on Amazon sharing your thoughts on *Full Stack Django and React*.

Go to the link below or scan the QR code to leave your review:

`https://packt.link/r/1803242973`

Your review will help us to understand what's worked well in this book, and what could be improved upon for future editions, so it really is appreciated.

Best Wishes,

Kolawole Mangabo

Download a free PDF copy of this book

Thanks for purchasing this book!

Do you like to read on the go but are unable to carry your print books everywhere? Is your eBook purchase not compatible with the device of your choice?

Don't worry, now with every Packt book you get a DRM-free PDF version of that book at no cost.

Read anywhere, any place, on any device. Search, copy, and paste code from your favorite technical books directly into your application.

The perks don't stop there, you can get exclusive access to discounts, newsletters, and great free content in your inbox daily

Follow these simple steps to get the benefits:

1. Scan the QR code or visit the link below

`https://packt.link/free-ebook/9781803242972`

2. Submit your proof of purchase
3. That's it! We'll send your free PDF and other benefits to your email directly

Printed in Poland
by Amazon Fulfillment
Poland Sp. z o.o., Wrocław

20609085R00246